Validation of the Measurement Process

James R. DeVoe, EDITOR
Institute for Materials Research,
National Bureau of Standards

A symposium sponsored by the Division of Analytical Chemistry at the 171st Meeting of the American Chemical Society, New York, NY, April 5–6, 1976.

ACS SYMPOSIUM SERIES **63**

AMERICAN CHEMICAL SOCIETY
WASHINGTON, D. C. 1977

Library of Congress CIP Data

Validation of the measurement process.
(ACS symposium series; 63 ISSN 0097-6156)

Includes bibliographies and index.

1. Chemistry, Analytic—Statistical methods—Congresses.
I. DeVoe, James R. II. American Chemical Society. Division of Analytical Chemistry. III. Series: American Chemical Society. ACS symposium series; 63.

QD75.4.S8V34 543'.01'82 77-15555
ISBN 0-8412-0396-2 ACSMC8 63 1–207 1977

PRINTED IN THE UNITED STATES OF AMERICA

ACS Symposium Series

Robert F. Gould, *Editor*

FOREWORD

The ACS SYMPOSIUM SERIES was founded in 1974 to provide a medium for publishing symposia quickly in book form. The format of the SERIES parallels that of the continuing ADVANCES IN CHEMISTRY SERIES except that in order to save time the papers are not typeset but are reproduced as they are submitted by the authors in camera-ready form. As a further means of saving time, the papers are not edited or reviewed except by the symposium chairman, who becomes editor of the book. Papers published in the ACS SYMPOSIUM SERIES are original contributions not published elsewhere in whole or major part and include reports of research as well as reviews since symposia may embrace both types of presentation.

CONTENTS

PREFACE

The existence of integrated electronic circuits has changed radically our thinking with respect to performing chemical analyses. Low cost microprocessors are now integral parts of commercial analytical instrumentation. Minicomputers have the ability to control experiments, to collect data, and to perform calculations with ever increasing facility. Thus, there is considerable interest on the part of the chemical analyst to use computational techniques to validate the measurement process.

Chapters 1 and 2 describe the importance of establishing statistical control of the measurement process and emphasize the use of graphical techniques which can be implemented conveniently on digital computers. After control of the measurement process has been established, it is necessary to evaluate systematic errors; Chapters 3 and 4 are devoted to this subject. Chapter 5 describes an innovative procedure which uses a laboratory minicomputer to optimize the variables in a chemical analysis. Chapter 6 outlines some examples for evaluating statistical control in testing laboratories.

I would like to thank the authors for their diligent effort and to express appreciation to Carol Shipley and the text editing staffs of the Analytical Chemistry Division and the Institute for Materials Research, NBS, for helping with the manuscripts.

Institute for Materials Research, NBS
Washington, DC 20234
August 12, 1977

JAMES R. DEVOE

PREFACE

1

Statistical Control of Measurement Processes

GRANT WERNIMONT

Department of Chemistry, Purdue University, Lafayette, IN 47905

Valid measurements are necessary whenever we make chemical tests on samples of materials so that proper action can be taken with regard to the bulk of the material. Measurements are not valid until we evaluate the performance characteristics of the process which produced the measurements and it is essential that the statements about the future behavior of these characteristics be correct. Statistical control is concerned with removing the assignable causes of variation in a measurement process (or correcting for their effects) so that we can associate approximate levels of confidence with these statements.

It is unfortunate, I think, that most academic courses involving measurement do not seem to make the student aware of how important it is to achieve a state of statistical control when we set up and run a measurement process. I was able to find only one current text on the theory and practice of quantitative analysis which addressed itself to this most important performance characteristic. In contrast, applied analytical chemists have been involved in statistical control activities for more than 40 years.

Some of the United States Government regulatory agencies are now becoming concerned about this important aspect of measurement operations. For example, the Nuclear Regulatory Commission requires (1):

> "The licensee shall establish and maintain a statistical control system including control charts and formal statistical

> procedures, designed to monitor the quality of each type of program measurement. Control chart limits shall be established to be equivalent to levels of significance of 0.05 and 0.001. When ever control data exceed the 0.05 control limits, the licensee shall investigate the condition and take corrective action in a timely manner. The results of these investigations and actions shall be recorded. When ever the control data exceed the 0.001 control limits, the measurement system which generated the data shall not be used for control limits, the measurement system which generated the data shall not be used for control purposes until the deficiency has been brought into control at the 0.05 level."

In this chapter the meaning of statistical control is explained, and the procedures which we can use to help set up and run a measurement process are reviewed so that it is in a state of statistical control.

WHAT IS MEASUREMENT

Measurement has been defined as "the operation of assigning numbers to represent properties using arbitrary rules based on scientific principles. Of course this is an over-simplification; a much broader interpretation of measurement formulates a hierarchy of measurement scales: Nominal, Ordinal, Interval, and Ratio (2). The mathematical transformations permitted on each scale determine what statistical methodology can be applied to the measurements. In general, the more unrestricted the permissable transformations, the more restricted the statistics; nearly all methodologies can be applied to ratio-scale measurements, but only a few serve for measurements on a nominal scale.

The most penetrating analysis, by far, of the basis for making measurements was formulated by Churchill Eisenhart (3); and it should be carefully studied by all people who devise measurement methods and perform measurement operations as well as by those who use measurement results to make decisions. Eisenhart states (3, p.163):

> "Measurement is the assignment of numbers to material things to represent the relations existing among them with respect to particular properties. The number assigned to some particular property serves to represent the relative amounts of this property associated with the object concerned.
>
> Measurement always pertains to properties of things not to the things themselves. Thus we cannot measure a meter bar, but can, and usually do measure its length; and we could also measure its mass, its density, and perhaps, also its hardness.
>
> The object of measurement is two fold: first, symbolic representation of properties of things as a basis for conceptual analysis; and second, to effect the representation in a form amenable to the powerful tools of mathematical analysis. The decisive feature is symbolic representation of properties, for which end numerals are not the usable symbols."

There is a form of direct measurement which is independent of the prior knowledge of any other property; but the number system used to express magnitudes must behave like the property being measured. A simple example of direct measurement is the use of JOHANSON blocks to calibrate a micrometer. In this case it is evident that the property we call length does behave like numbers in the following two ways:

1. An experimental procedure can be devised which will produce an ordered sequence of the blocks.

2. Another experimental procedure can be devised to combine (wring) the blocks additively.

A more complex example is the property we call absorbance (A = -log Transmittance) which behaves according to the rules of matrix algebra (4).

In analytical chemistry, measurements are occasionally made by a direct method; but for reasons of convenience, we more often use an indirect method based on fundamental or empirical laws involving various physical, chemical and biological properties of matter, energy and radiation.

It is a fact of experience that a measurement process can be used under more diversified conditions when it is based on fundamental laws rather than on empirical laws for which we have inadequate theoretical explanations. However, we should never forget that empirical laws are involved in almost every measurement process.

MEASUREMENT METHODS AND PROCESSES

Let us now examine how we devise and carry out the operations to make measurements. Eisenhart has discussed this in great detail (3, p. 165) and I extract some of his points:

> "Specification of the apparatus and equipment to be used, the operations to be formed, the conditions in which they are carried out - these instructions serve to define a method of measurement...
>
> A measurement process is the realization of a method of measurement in terms of a particular apparatus, equipment, conditions, etc. that at best only approximate those prescribed...
>
> Written specifications in methods of measurement often contain absolutely precise instructions which cannot be carried out (repeatedly) with exactitude in practice...to this extent there are certain discrepancies between a method and its realization by a particular process...
>
> The specification often includes imprecise instructions such as 'raise the temperature slowly', 'stir well', etc...to the extent the instructions are not absolutely definite, there will be room for differences between realizations of the same measurement method...
>
> To qualify as a specification, a set of instructions must be sufficiently definite to insure statistical stability of repeated measurements of a single quantity,

that is the measurement process must be capable of meeting the criteria of statistical control."

Of course, we must carry out a well designed series of experiments to devise the types of equipment and the sequence of operations which make up the measurement process before we can formulate the measurement method; and it is necessary that we find optimum conditions for running the process so that it responds to significant changes in the level of the property being measured, but does not respond to small changes in its operating conditions (5). My experience leads me to conclude that we fail to recognize how difficult it is to (a) find the optimum conditions, (b) control all the significant assignable causes of variation, and (c) write a concise yet unambiguous set of procedural instructions.

Measurement processes in chemical analysis consist of unit operations which include,

- taking a gross sample from an aggregate of material,
- taking a laboratory subsample from the gross sample,
- treating the subsample, physically and chemically to remove interferences,
- measuring a property of the treated subsample, and
- estimating the desired property using a calibration curve.

A useful technique for visually showing the measurement operations is a block diagram or flow chart. The person who developed the process, or has run it repeatedly, can easily draw the flow chart; it will supplement the method and help other poeple to understand the process.

It is important to realize that the final measurement is an attribute of the laboratory sample, and is only an estimate of the property in the entire aggregate of material. We must never forget that we make inferences about the magnitude of the property in the aggregate from a small finite group of measurements on the laboratory sample.

If we fail to control the significant assignable causes which can, potentially, affect the various operations in the measurement process, we will certainly find that it will not meet the criteria of statistical control.

STATISTICAL CONTROL

I am unable to give a simple, concise definition of what we mean by "the state of being in statistical control". The concept was conceived and developed by Dr. Walter A. Shewhart, a physicist-engineer at the Bell Telephone Laboratories, to help solve the problems of manufacturing products of uniform and acceptable quality. While his publications (6,7,8) are not primarily concerned with measurement processes, they do present ideas which can be applied to them. The 1939 book gives a philosophical discussion of statistical control, the presentation of measurement results, and the specification of precision and accuracy.

Eisenhart presents a section of the requirement of statistical control (3, p. 166) which summarizes Shewhart's ideas and demonstrates how they apply to measurement processes; I extract some of these ideas:

> "The point that Shewhart makes forcefully, and stresses repeatedly, is that the first *n* measurements of a quantity generated by a measurement process provide a logical basis for predicting the behavior of further measurements of the same quantity by the same measurement process, if and only if, these *n* measurements may be regarded as a *random* sample from a 'population or universe' of all conceivable measurements...characterized by a probability distribution...nothing is said about the mathematical form of the distribution. The important thing is that there be one...
>
> Shewhart was well aware that, from a set of *n* measurements in hand, it is not possible to decide, with certainty, whether they do or do not constitute a random sample from some definite statistical population characterized by a probability distribution. He therefor proposed (7) that in any particular instance one should 'decide to act for the present as if' the measurements in hand (and their immediate

successors)...meet the requirements of the small sample version of Criterion I of his previous book (6) and...show no evidence of lack of statistical control when analyzed for randomness in the order in which they were taken by the control chart techniques, for averages and standard deviations that he had found so valuable in industrial process control, and by certain additional tests for randomness based on 'runs above and below average and runs up and down'...

Experience shows that in the case of measurement processes, the idea of strict statistical control that Shewhart prescribes, is usually very difficult to attain, just as in the case of industrial production processes..."

Eisenhart also quotes from a paper by Dr. R. B. Murphy, another Bell Telephone engineer, on the validity of precision and accuracy statements (9):

"...a test method ought not to be known as a measurement process unless it is capable of statistical control...(which) means that either the measurements are the product of an identifiable statistical universe, or if not, the physical causes preventing such identification may themselves be identified and, if desired, isolated and suppressed. Incapability of control implies that the results of the measurement process are not to be trusted as indications of the property at hand - in short, we are not in any verifiable sense measuring anything...without this limitation on the notion of a measurement process, one is unable to go on to give meaning to those statistical measures which are the basis for any discussion of precision and accuracy."

I believe we can now formulate the idea of statistical control as follows: A measurement process may be said to be in a state of statistical control if the significant assignable causes of variation have been removed or corrected for, so that a finite set of *n* measurements from the process can be used to

(a) predict limits of variation for the n measurements and (b) assign a level of confidence that future measurements will lie within these limits.

CONTROL CHART ANALYSIS

The operational procedure for demonstrating that a process is in a state of statistical control is quite simple in concept but rather complex in practice. It consists of arranging to gather n measurements, in some kind of order, and in the form of so-called "rational subgroups", within which the variations may be considered, on the basis of a knowledge of the process, to be random causes only, but between which, the variations may be due to suspected assignable causes.

To illustrate how we make a control-chart-analysis of measurements, let us examine the results of a simple experiment which Shewhart carried out to simulate a "controlled" production process. He placed 998 circular chips in a large bowl; numbers between negative 3.0 and positive 3.0, at 0.1 intervals, were recorded on the chips which were one color for the negative numbers and another for the positive. The magnitudes of the numbers were distributed according to a "normal" distribution with average = 0.0 and standard deviation = 1.007. The chips were drawn from the bowl one at a time, with replacement, until 4000 values were obtained and recorded in order. For further details, see (6, pp. 164-165 and Appendix II).

Shewhart observed that in this experiment we have as near an approach as is likely feasible to the conditions in which the law of large numbers applies since, to the best of our knowledge, the same essential conditions were maintained. However, he once told me that this simple drawing operation is prone to show lack of statistical control unless great care is taken to mix up the bowl of chips between the drawings and keep the bookkeeping mistake-free.

I have plotted the results of the first 200 drawings as a control chart in Figure 1, using a rational subgroup of four consecutive values. The averages and standard deviations of the 50 subgroups were calculated as,

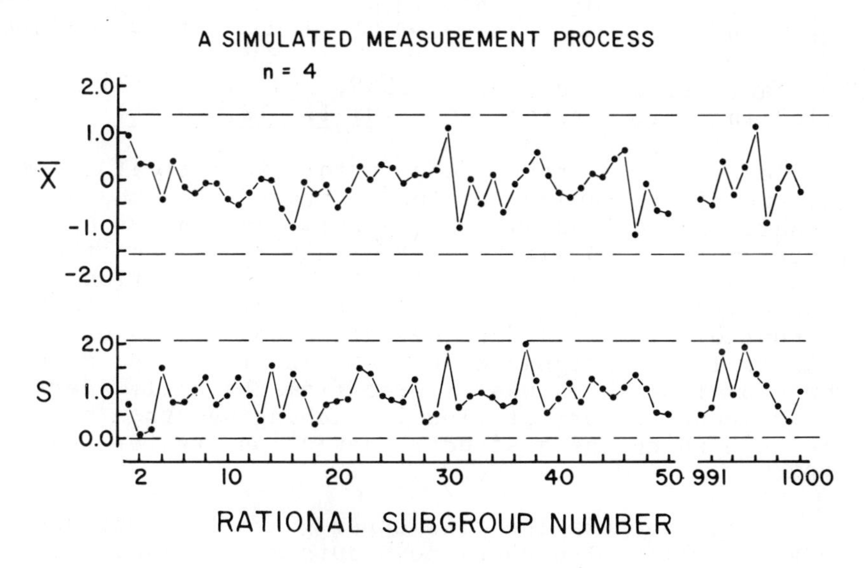

Figure 1. Consecutive drawings from Shewhart's bowl of chips

$$\bar{X} = \Sigma X_i/4, \text{ and } S = \sqrt{\Sigma(X_i-\bar{X})^2/(4-1)}.$$

The grand average, of all 200 values is -0.08 and the average of the group standard deviations is 0.912. Three-sigma control limits for the 50 subgroups are,

Limits	Standard Deviation	Average
Upper	(2.266 x 0.912) = 2.07	-0.08 + (1.628 x 0.912) = 1.40
Lower	(0. x 0.912) = 0.	-0.08 - (1.628 x 0.912) = -1.40

The factors, $B_3 = 0_o$, $B_4 = 2.266$, and $A_3 = 1.628$ are tabled in various references (10, 11, 12, 13, 14).

To evaluate these results for statistical control, we first examine the lower graph of subgroup standard deviations which, in this case, shows none greater than the 3-sigma limit. This indicates that no assignable causes were affecting the operation of consecutively drawing and replacing four chips. Lack of control for standard deviation would lead us to look for local assignable causes in the way each group of four chips was removed from the bowl. Perhaps someone is surreptitiously exchanging the bowl with one which has a standard deviation greater than 1.007.

Next, we examine the upper graph for subgroup averages, which also shows none outside 3-sigma limits. This indicates that no assignable causes were affecting the drawing operation through out the entire sequency of the first 200 values. Lack of control would suggest that some nonlocal assignable cause affected some subgroups differently than others. Perhaps the surreptitious exchange involved a bowl with a distribution which averages two rather than zero.

Shewhart suggested that criteria for randomness should also include the behavior of urns for consecutive groups within the 3-sigma limits. Duncan explains (13, p. 386) a run as "a succession of items of the same class" such as a series of increasing or decreasing value, or a series of consecutive values above or below the average. We find no runs, up or down, greater than five; but two runs, of seven below the average, occurred (beginning with subgroups 6 and 15). Statistical theory and practical experience indicate that assignable causes can usually be found to explain runs of seven or more; of course it is now impossible to look for them.

No other types of systematic variation such as cycles or trends, appear to be present for either the standard deviations or the averages. Can we conclude that this process was in a state of statistical control? Well, we have two choices: (a) the process was not in control, or (b) the process was in control but two improbable runs occurred. This is exactly the situation we meet almost every time we examine results from a measurement process. No matter which choice we make, there is some chance that it is wrong. I would conclude that the evidence for lack of control is not convincing based on knowledge of the process, and predict that the 3-sigma limits, estimated from the first 200 drawings, should also include practically all of the remaining 3800. You can see that the last 40 drawings from Shewhart's bowl are well within these limits.

Duncan has given (13, p. 392) the following summary of criteria for lack of statistical control:

1. One or more points outside 3-sigma limits,
2. One or more points in the vicinity of a "warning limit" suggesting that additional observations be taken,
3. A run of seven or more points,
4. Cycles, trends, or other nonrandom patterns within 3-sigma limits,
5. A run of two or more points outside of 2-sigma limits,
6. A run of four or more points outside 1-sigma limits.

Of course we are always faced with the risk of being wrong when we decide whether, or not, a process is in a state of statistical control. We fix this risk by arbitrarily choosing critical 3-sigma limits. Using wider limits, we increase the risk of erroneously concluding that the process is in statistical control and decrease the chances of detecting significant assignable causes. The use of narrower limits will have the opposite effects. Experience has shown that the risks are quite tolerable, in most cases, when action limits are set between 2- and 3-sigma for subgroup standard deviations and averages.

RATIONAL SUBGROUPS

The key to success when we use control chart analysis to examine results from a measurement process, lies in the strategy we use to set up "rational" subgroups. The idea of arranging to gather the measurements in subgroups makes real sense, because it is my observation that assignable causes affecting a measurement process fall rather clearly into two classes.

The first class is under the <u>local</u> control of the person who operates the process; it includes such operations as manipulating equipment, dispensing reagents, calibrating instruments, judging indicator end points, and otherwise following procedural instructions in local time and space. Operators can be held responsible for maintaining rigid control of these local operations, and good operators soon learn how to do it. Lack of statistical control of these local operations is observed, occasionally, but only because of basic shortcomings in the method or equipment which the operator is unable to perceive or cope with.

The second class of assignable causes is not under the local control of the operator; it includes such things as long-range maintainance of laboratory conditions and equipment, types and/or methods of calibration, deterioration of reagents and instruments, the nature of interferences in the material being tested, and numerous other types of nonlocal or regional assignable causes. The laboratory supervisor must assume responsibility for finding and removing assignable causes affecting these operations.

I think it is obvious that control chart analysis for variation within rational subgroups (standard deviation or range) gives us important information about the local assignable causes, while the chart for averages reveals information about the regional assignable causes.

Two possible mistakes are easy to make when we set up a system of rational subgroups: (a) the replications are so close together in time and/or space that they do not include all the local assignable causes. For instance, we would never want to record duplicate readings of an instrument scale because, as W. J. Youden often pointed out, this is merely "duplicity". The subgroup should include all the local random causes because a measurement process can never be brought into a state of statistical control if the rational subgroups are too restricted, (b) the replications are so far apart in time and/or space that they include some of the regional assignable causes. This leads to wide control limits which lack the power to detect assignable causes, local or regional.

I have a detailed discussion of the concept of rational subgroups in my paper, "The Use of Control Charts in the Analytical Laboratory" (15). Specific instructions cannot be formulated to devise rational subgroups which will apply to all kinds of measurement processes. In general, the subgroups should be limited so that variations within the subgroups is essentially random and they should be sufficiently extended to reveal assignable causes which the operator is unable to control.

Let us now look at some real world examples of how we can use control chart analysis.

A PROCESS WITH NO ASSIGNABLE CAUSES

Figure 2 shows a control chart for a process to determine the water-equivalent of a Parr-type bomb combustion calorimeter. Once each month, the operator made four independent calibration runs on the same afternoon by weighing appropriate amounts of NBS Standard Benzoic Acid and burning it in the oxygen-charged bomb under essentially the same conditions as were used to determine heats of combustion of fuel. The material was ignited by heating electrically a small piece of pure iron wire. The calorimeter constant was computed from the observed temperature rise of the water surrounding the bomb, the weight of benzoic acid, and the NBS certified value for the heat of combustion of the acid. A small correction for the heat generated by the wire was applied.

The data for this chart was taken from historical records and you can see that during the previous 11-month period, no significant assignable causes were affecting the standard deviations so we can conclude that the operator was controlling all the local operations. The chart for averages also shows satisfactory control which means that no regional assignable causes were affecting the calibration operations over an extended period of time.

It is interesting to note that prior to this analysis of calibration measurements, the laboratory supervisor had been revising the water-equivalent each month. He now decided to adopt the long range average of 29030 but continue checking it every month as before. This was sound strategy because a few months later the calibration average was observed to be just out of control on the low side. Investigation revealed that a new supply of iron wire had been acquired but the supervisor neglected to give a revised correction factor to the operator.

A PROCESS WITH LOCAL ASSIGNABLE CAUSES

I have already indicated that lack of control of local assignable causes is not commonly observed; and I am aware of no simple techniques, other than control chart analysis, to detect it. This example involved the use of an instrument to measure the tearing strength of plastic sheeting used to support photographic emulsions. The instrument (Thwing-Albert), designed to measure the tearing strength of paper, consisted of a fairly massive pendulum arranged so that it could absorb the energy used to tear a small specimen of material, thus decreasing the amplitude of the pendulum.

The instrument had been modified to make it more sensitive to the smaller strengths of film support by attaching a counterbalance to the pendulum, thus raising its center of gravity. The modified instrument was monitored by means of a reservoir of "reference" film support picked from a uniform production lot, cut into test specimens, and thoroughly randomized. The specimens were conditioned and tearing strengths were measured once each day using rational subgroups of five strips from the reservoir.

Control chart analysis showed no evidence for lack of statistical control for both standard deviation and average during the first 14 weeks as you can see in Figure 3. During week 20, lack of control was indicated for one subgroup standard deviation and one average; and by week 24, it became evident that both standard deviation and average were out of statistical control. The operator could find no reasons to

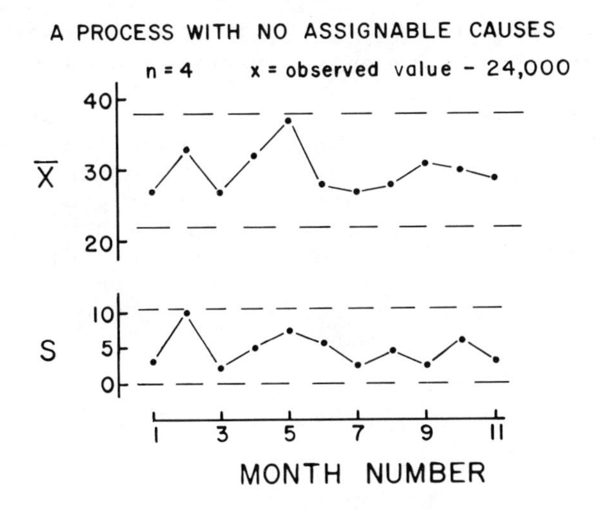

Figure 2. Determination of the water-equivalent of a bomb calorimeter

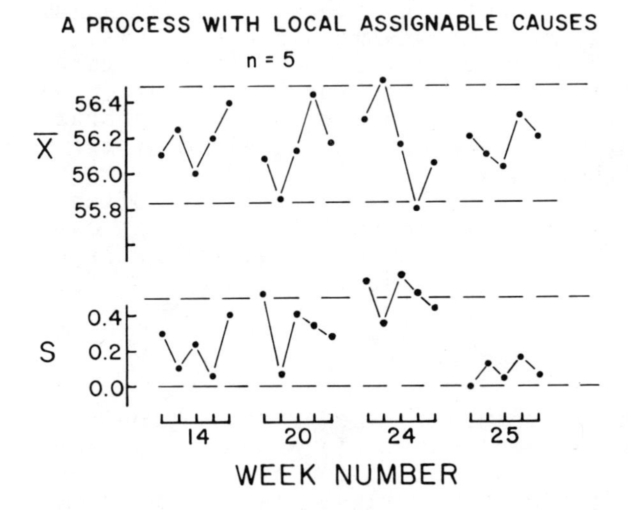

Figure 3. Determination of the force to tear plastic film support

explain this and the material behaved alright when tested on other instruments currently in use.

The instrument was returned to the machine shop where the counterbalance had been installed, and it was found that the bearings, on which the pendulum was supported, were beginning to disintigrate because of the increased load of the counterbalance. Larger bearings were installed and, as you can see, the control chart for both standard deviation and average returned to normal. New bearings had to be installed on all the other instruments.

A PROCESS WITH REGIONAL ASSIGNABLE CAUSES

When control chart analysis shows satisfactory control for the variation within rational subgroups but lack of control among subgroup averages, we must look for regional assignable causes. Most interlaboratory studies of measurement processes show little or no evidence for lack of control within the laboratories over a short period of time; but it is very difficult to achieve statistical control among a group of laboratories all using the same test method. Figure 4 shows results of a study of the Eberstadt method for determining the acetyl-content of cellulose acetate. Samples of a reference material were analyzed in eight different laboratories with two independent operators in each laboratory making duplicate tests on each of two different days. The lower chart for operator ranges shows that a state of statistical control existed for the variation within the laboratories, but it is obvious that laboratory averages vary more than can be explained by the variation within laboratories. It is difficult to find the reasons for this because they are often different from one laboratory to another. In this case it was found that some of the laboratories were not rigorously following the test method procedures.

THE PROBLEM OF DUPLICITY

Let us return to the critical problem of devising rational subgroups. In Figure 5, we see results for the determination of copper, made during the production of bronze castings. Two independent samples were drilled from each casting and analyzed, in duplicate, using a precise method of electrolytically

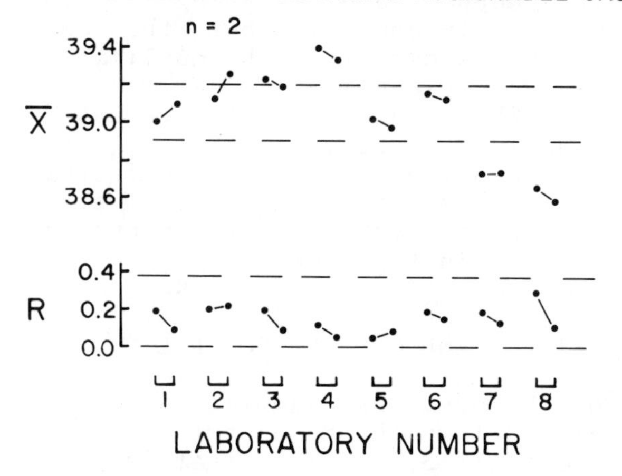

Figure 4. Determination of acetyl in cellulose acetate

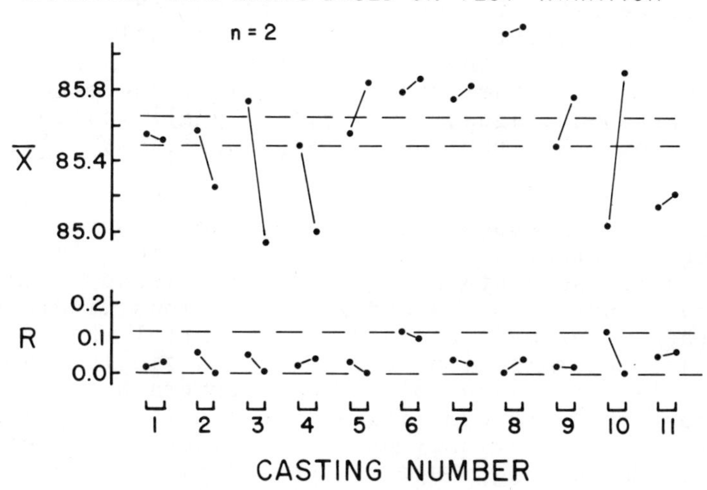

Figure 5. Determination of copper in bronze castings

depositing the copper and weighing it. The lower chart for ranges is in control, but the chart for subgroup averages shows that the duplicate samples are exceedingly variable compared to the duplicate determinations. When control limits are based on the variation of sample averages, within castings, there is some reason to believe that the manufacturing and testing operations are both in a state of statistical control, although a cyclic effect cannot be ruled out, as you can see in Figure 6.

SIMPLE AND COMPLEX CONTROL

In all of the previous examples, we have assumed a mathematical model which Eisenhart called SIMPLE statistical control (3, p. 174), that is, the variation of measurements within rational subgroups is random and serves as a valid estimate of the random variation of the subgroup averages. However, we often find processes for which this model is inadequate because regional assignable causes exist which we cannot identify and/or remove; in such cases, it is desirable to determine whether the process is in a state of COMPLEX, or multistage, statistical control (3, p. 178).

We do this by setting up a control chart for the variation (standard deviation or range) of measurements within the rational subgroups, just as before. However, we estimate control limits for the subgroup averages by treating them as "individual" measurements and then use the "moving range" method which calculates all the consecutive differences between the subgroup averages, thus partially eliminating the effects of the regional assignable causes (13, p. 451).

Figure 7 shows results for the measurement of the water content of a series of production lots of an organic solvent using the Karl Fischer method. The lower chart for standard deviations indicates that the measurement process is in control when three replicate determinations are made on a single sample of material from each lot. The upper graph shows the averages; the narrow limits are based on replicate measurement variation, while the wide limits correspond to the moving range of consecutive lot averages. Of course, we would not expect the distillation of an organic material to be in simple statisti-

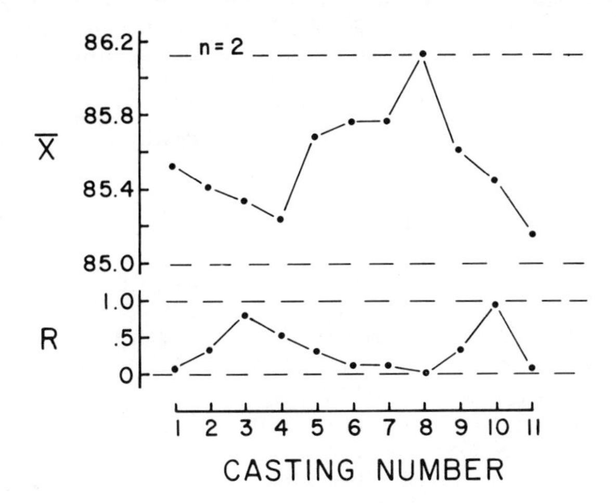

Figure 6. Determination of copper in bronze castings

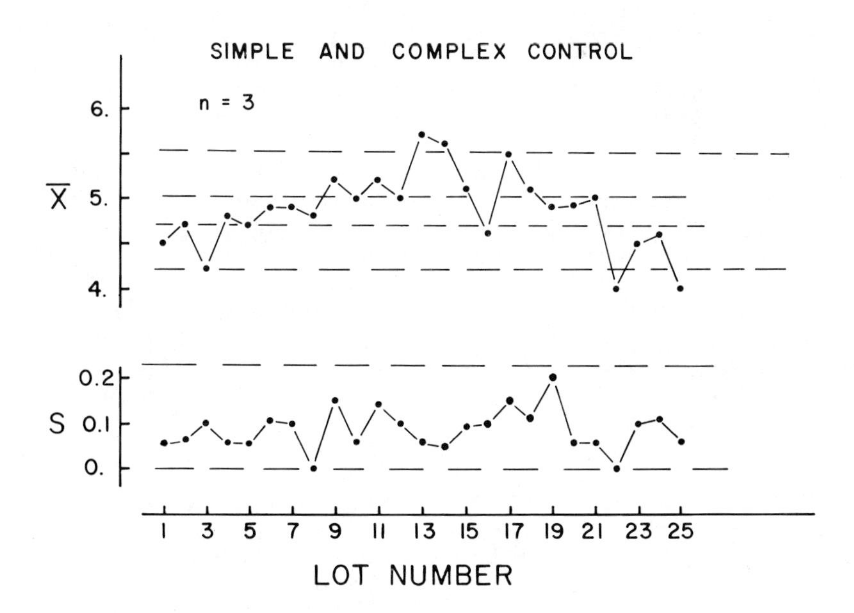

Figure 7. Determination of water in an organic solvent

cal control in a case like this; but we see that the overall operation of distilling and measuring is not even in the state of complex statistical control. The assignable causes for this may be in the measurement of the water content, but they are much more likely to be found in the distillation process.

UNORDERED DATA ANALYSIS

Control chart analysis was originally applied to measurements taken in sequential order from a continuous process, but it can also be used to compare results from different sources where logical order cannot be assigned. An important example is the interlaboratory study of a test method (16, 17). Here it is necessary to give very serious thought of how to arrange for subgroups within the laboratories. Some people have defined a subgroup as the measurements made by a single operator, using a single set of equipment, as closely together as possible. This can be considered to be duplicity. A more useful subgroup includes the local assignable causes over a more reasonable period of time, for example, a week or more. A logical reason for this more extensive rational subgroup is the fact that the people who use measurement results, often require comparisons between repeated measurements to help make decisions relating to sample rechecks, production changes, material sources, etc., made over the interval of this period of time.

Many of these control chart methods were developed by Shewhart and successfully used by many people for nearly fifty years. During the last three decades, more sophisticated control charts for such things as cumulative sums, lot acceptance, multivariable responses, etc., have been developed (18); and some of these techniques will be found useful to help evaluate measurement processes.

RELATED ASSIGNABLE CAUSES

Many measurement processes show lack of statistical control of a type which often appears baffling because the assignable causes act together so that the effects of one are not the same at various levels of the other. For example, it has long been known that the oxidation of ferrous iron with potassium permanganate gives high results in hydrochloric acid solutions; the deviations increase with acid concentration. Also, the deviations are relatively smaller as the iron concentration increases, and the rate of titration decreases. It is most important that we find and remove the effects of this kind of differential response while a measurement process is being developed.

The classical experimental procedure (sometimes called the scientific method) for optimizing the response of measurement process is inadequate to detect this kind of related behavior between assignable causes. In the case of two factors, such a procedure studies each, at some fixed level of the other as is shown in Figure 8 on the left; but it never determines whether the effects of changing the levels of the factors are independent of each-other. Differential response is easily detected using a complete factorial design as is shown on the right, where the effects of all combinations of the factors are measured with little or no extra work. In this case, the factors are acting independently if the difference between the diagonal averages is not significantly greater than zero.

Differential response (usually called interaction, or nonadditivity by statisticians) can be of three types: (a) among factors within the measurement process, (b) between process factors and the type of material being tested, and (c) between test methods and the type of interferences in the material being tested.

The example described above falls into the first type. Figure 9 shows the problem of differential response when several materials are tested using a measurement process set up in various laboratories. The laboratories do not rank the materials in exactly the same order. This behavior is not serious as long as the variation among the laboratories is no greater than the replication error of the process. However, when unknown interferences are present in different types of material, which affect some laboratory results but not others, it soon becomes impossible to predict the response of the test method on types of material, other than those used in the interlaboratory study.

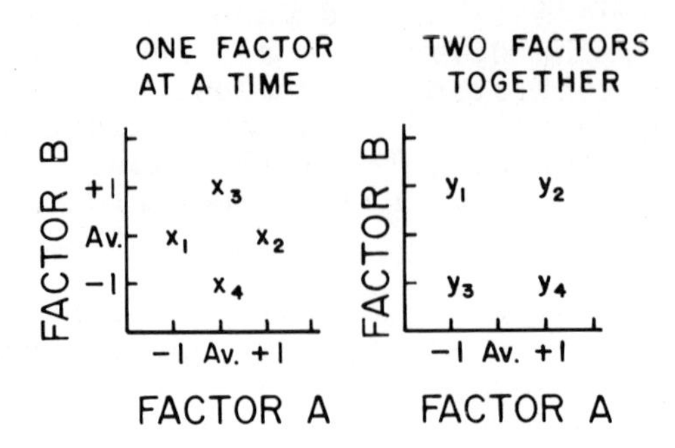

Figure 8. Two types of experimental designs

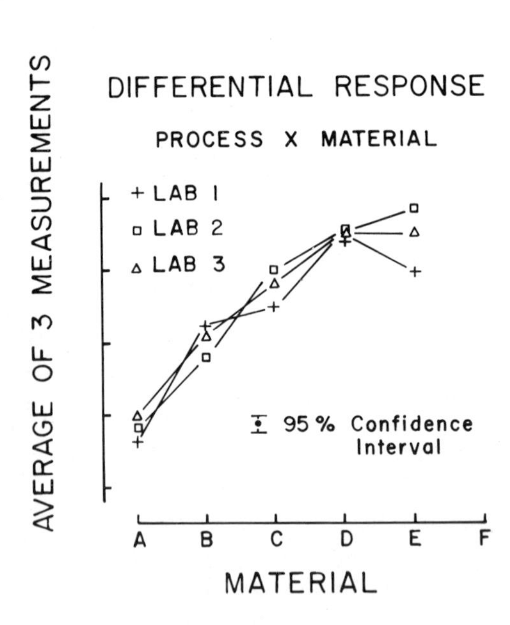

Figure 9. Resistance of floor materials to surface abrasion

Differential response is often observed when two test methods are based on slightly different physical and chemical principles. This has been observed for methods to evaluate material flammability, surface abrasion, fabric wrinkle resistance, paper smoothness, etc. We are usually unable to explain these interactions, especially if the fundamental principles of the methods are only vaguely understood.

RUGGEDNESS OF A MEASUREMENT PROCESS

Measurement processes are often developed in a single laboratory and then used in other laboratories. We have already noted that the variation of results from different laboratories is usually not in a state of statistical control with respect to the variation within laboratories. Occasionally we are able to identify some of the assignable causes but only with a great deal of effort.

Dr. W. J. Youden addressed himself to this important problem and he observed (19):

> "By no means an unusual occurrence is a collaborative test whose results obviously fall short of expectations based on data from the initiating laboratory. The explanation is usually found in the fact that the initiating laboratory has a set of operations and equipment that is never varied. In fact, care is taken not to vary the routine in any particular. Naturally no light is shed on what may happen when the procedure on trial is used by a number of laboratories each of which establishes its own particular routine".

He goes on to suggest that things like the source, age and concentration of reagents, the rate of heating solutions, the temperature and time of drying materials, the environmental conditions of temperature and relative humidity, and many other factors may not be specified in detail so that they vary, within small limits, from one laboratory to another.

The only protection against this type of assignable cause is for the initiating laboratory to deliberately introduce minor variations into the procedure and observe what happens. At first this appears to involve a great deal of extra work, most of which may yield negative results. Youden suggested a scheme of attack that will conserve labor yet be sensitive enough to pick up fairly small effects.

The principle of his suggested design can be illustrated by a simple experiment involving just three factors (20). Let the levels of the factors, chosen slightly above and below the specified operating conditions, be designated A, B, C and a, b, c respectively. Only four experiments need to be run :

Run Number	Factor Level	Observed Result
1	ABC	t
2	aBc	x
3	abC	y
4	Abc	z

Notice that two factors are always changed from one run to another. The differential effect of changing the level of each factor is obtained by computing the averages,

$$\bar{A} = (t+z)/2;\ \bar{B} = (t+x)/2;\ \bar{C} = (t+y)/2,$$

$$\bar{a} = (x+y)/2;\ \bar{b} = (y+z)/2;\ \bar{c} = (x+z)/2.$$

The two runs for $\bar{A}$ involve the levels B, b, C, and c for the other two factors and this is also true for $\bar{a}$; thus if $\bar{A}$ differs significantly from $\bar{a}$, the first factor must be the assignable cause. The same logic applies to the other two factors.

It is important to recognize that the justification for this logic rests on the expectation that changes in the levels of all the factors has been quite small and are not supposed to have appreciable effect on the measurement process.

This experimental design has many advantages over the traditional one-factor-at-a-time scheme:

1. The number of experiments is minimized at one more than the number of factors being studied, although only certain combinations are possible.

2. Differences are evaluated in terms of averages so that the discriminating power is greater for the same number of runs.
3. If no significant factors are found, we can get a preliminary estimate of how the method will behave in other laboratories.
4. If significant factors are found, their effects can be estimated and appropriate tolerances set for their control.

The restriction to certain combinations of factors and runs is not very serious; Youden considered the plan for seven factors in eight runs to be a good compromise and he has published several examples (20, 21). Other people have reported results of experiments to test the ruggedness of a test method and, without exception, they were able to detect one or more potential assignable causes of variation in the test method, as written (22).

A MEASUREMENT HIERARCHY

Experience tells us that some measurement processes can be easily brought into a state of statistical control while others seem subject to a plethora of assignable causes that are hard to locate and difficult to control. Why is this so?

Let us arrange measurement operations into the following order:

1. Measurement processes to determine natural constants,
2. Calibration of physical and chemical reference processes,
3. Precise and accurate "standard" measurement processes,
4. Routine control measurement processes, and
5. Laboratory simulation processes to measure performance characteristics.

The determination of natural constants such as the speed of light, the acceleration of gravity, the atomic weights of elements, etc., requires that we spare no effort to correctly assign a precise and unbiased value to represent the property involved.

Operations to calibrate weights, volumetric-ware etc., and to prepare pure chemical compounds and assign values to homogeneous reference materials, fall into the second class.

The third class includes "standard" measurement processes which are usually fairly complex so that all significant interferences are removed or corrected for.

Routine control measurements are in the fourth class; they require less elaborate equipment than the standard methods and are more economical to run although they give less precise and less accurate results.

Finally, the last class involves measuring the performance of a system rather than the magnitude of a property. Resistance to surface abrasion or weathering, flammability of children's sleep-wear, the efficiency of removing dirt from a carpet, are examples of this class.

You will surely recognize several types of order as we proceed from top to bottom in this hierarchy:

-the scientific principles involved, are fundamental and fairly well understood at the top; they are empirical and very complex at the bottom,

-operations to correct for the effects of known interferences become increasingly more difficult,

-assignable causes are more difficult to identify,

-differential response becomes more prevalent and more difficult to cope with, and

-ruggedness against stresses on the operational procedure decreases as we go down the hierarchy.

We should keep this classification in mind when ever we devise and develop a measurement process; its position in the hierarchy is an indication of the problems we must solve in order to maintain the process in a state of statistical control.

SUMMARY

Measurements are not valid until we evaluate the performance characteristics of the process which produces them. Statistical control is concerned with removing all significant assignable causes which can affect the process so that statements about the variability of future results will be correct with an associated level of confidence.

Control chart analysis, first developed by W. A. Shewhart, is used to examine a finite sequence of statistical control. The measurements are divided into "rational subgroups" and the standard deviation, within subgroups, serves to estimate limits for the variation of the subgroup-standard deviations and - averages. Lack of statistical control is indicated when these limits are exceeded or when nonrandom patterns of variation occur within the limits.

The concept of rational subgroups makes sense because assignable causes of process variation fall rather clearly into two classes: local manipulations which are under the control of the operator; and regional operations, in time and space, to maintain the stability of the process, for which someone, other than the operator, must be responsible. We observe a lack of statistical control for subgroup-averages more often than for standard deviations because regional assignable causes are difficult to find and remove.

Not infrequently, we find that two assignable causes act together so that their effects are not additive, that is, the effects of one are not the same at all levels of the other. Process interferences, also may be present in some types of material being tested and not in others. We find it difficult to understand this kind of process response unless we use complete factorial experimental designs.

Lack of statistical control among laboratories is inevitable unless the measurement method is rugged against small changes in process operating conditions. W. J. Youden has provided us with an efficient experimental design to test the ruggedness of a method.

Finally, we can arrange measurement operations into a hierarchy which clearly shows that the better we understand the scientific principles involved, the easier it is to maintain the process in a state of statistical control.

LITERATURE CITED

1. Federal Register (1975) Vol. 40, No. 155, p. 33653.
2. Stevens, S. S. in "Measurement, Definitions and Theories", C. W. Churchman and R. Philburn, Ed., Chapter 2, John Wiley and Sons, Inc., New York, 1959.
3. Eisenhart, Churchill, Journal of Research of the National Bureau of Standards -C. Engineering and Instrumentation (1963) Vol. 67C, No. 2, pp. 161 to 187.
4. Wernimont, Grant, Anal. Chem. (1967) Vol. 29, pp. 554-562.
5. Wernimont, Grant, Materials Research and Standards (1969) Vol. 9, No. 9, pp. 8-21.
6. Shewhart, W. A., "Economic Control of Quality of Manufactured Product", D. Van Nostrand Company, Inc., New York, 1931.
7. Shewhart, W. A., "Statistical Method from the Viewpoint of Quality Control", The Graduate School, U. S. Department of Agriculture, Washington, 1939.
8. Shewhart, W. A. in "Contribution of Statistics to the Science of Engineering, University of Pennsylvania Bicentennial Conference", University of Pennsylvania Press, 1941.
9. Murphy, R. B., Materials Research and Standards (1961) Vol. 1, No. 4, pp. 264-267.
10. "ASTM Manual on Quality Control of Materials", American Society for Testing and Materials, Philadelphia, 1976.
11. "Definitions, Symbols, Formulas, and Tables for Control Charts", American Society for Quality Control, Milwaukee, 1972.
12. "Glossary and Tables for Statistical Quality Control", American Society for Quality Control, Milwaukee, 1973.
13. Duncan, A. J., "Quality Control and Industrial Statistics, 4th Ed.", Richard D. Irwin, Inc., Homewood, 1974.
14. Bicking, C. A. and Gryna, F. M., Jr. in "Quality Control Handbook, 3rd. Ed.", J. M. Juran, Ed., Section 23, McGraw-Hill Book Company, New York, 1974.

15. Wernimont, Grant, Anal. Chem. (1946) Vol. 18, pp. 587-592.
16. Wernimont, Grant, ASTM Bulletin (1950) No. 166, pp. 45-48.
17. Wernimont, Grant, Anal. Chem. (1951) Vol. 23, pp. 1572-1576.
18. Gibra, I. N., J. Qual. Tech. (1975) Vol. 7, No. 4, pp. 183-192.
19. Youden, W. J., J. Assoc. Offic. Agr. Chem. (1963) Vol. 46, p. 56.
20. Youden, W. J., Materials Research and Standards (1961) Vol. 1, No. 11, p. 863.
21. Youden, W. J. and Steiner, E. H., "Statistical Manual of the AOAC", p. 50, The Association of Official Analytical Chemists, Washington, 1975.
22. Wernimont, Grant, in "Symposium on Preparation and Use of Precision Statements", American Society for Testing and Materials, Philadelphia, In Press.

2

Testing Basic Assumptions in the Measurement Process

JAMES J. FILLIBEN

National Bureau of Standards, Washington, DC 20234

The purpose of this paper is to discuss various statistical techniques for the testing of the basic underlying assumptions in a measurement process. In general, the measurement process refers to the act of collecting quantified information about some phenomenon of interest under well-defined conditions. Among the various components of the measurement process are the experimentalists themselves. The end objective in a measurement process is predictability (1,2) that is, the ability to make probability statements about measurements already taken and yet to be taken. If this predictability is not present, then the process will yield conclusions which are only temporal and local in nature, and which will lack the generality typical of scientific experimentation. To achieve such predictability, the measurement process must be "in control (3)." The term "in control" is a statistical term--having nothing to do *per se* with whatever physical science area or phenomenon that the experiment involves, but rather with the properties of sequences of measurements. A broad definition of "in control" is as follows:

> a measurement process is in control if the resulting observations from the process, when collected under any fixed experimental condition within the scope of the *a priori* well-defined conditions of the measurement process, behave like random drawings from some fixed probability distribution with fixed location and fixed variation parameters.

The essential components implied by the above definition are:

1. randomness

2. fixed location

3. fixed variation

4. fixed distribution

It is to be noted that "fixed location" as used above and throughout this paper is an abbreviated way of stating that the measurement process has a single limiting mean (i.e., as the measurement process continues in time, it is conceived to have a unique limiting "typical value"). Similarly, "fixed variation" is an abbreviated way of stating that the measurement process has a stable degree of variation. It is important of course to note that the above components in the definition of "in control" are identically those underlying assumptions which are typically made, either knowingly or unknowingly, in a measurement process. The consequences for invalidity of these assumptions are the arrival at incorrect conclusions and the loss of the desired-for predictability that the scientist invariably seeks.

In this light, the testing and checking of basic assumptions (5) in a measurement process takes on its rightful importance. The testing of assumptions is a "necessary evil," tangential in a sense to the main analysis, but which rarely can be short-circuited. In order to test such assumptions after the fact, we have only the raw data resulting from the experiment. Fortunately, however, much information about the validity or invalidity of the underlying assumptions is still latent in the data and considerable progress (of both a theoretical and practical nature) has been made in the last decade in the development of statistical techniques for the extraction of such information. The remainder of this paper will deal with an enumeration and discussion of such techniques.

It will be noted that almost all of the techniques to be presented are graphical in nature. There are many reasons for such a heavy dependence on graphics:

1. Plots take full advantage of the pattern recognition capabilities of the human (e.g., linearity is easy to detect).

2. Plots make use of a minimal number of assumptions. Thus, the use of plots increases the likelihood that the conclusions will not be approach-dependent.

3. From a communications point of view, a plot is generally a much more understandable and efficient way of conveying information to another analyst or experimentalist than is a set of statistics.

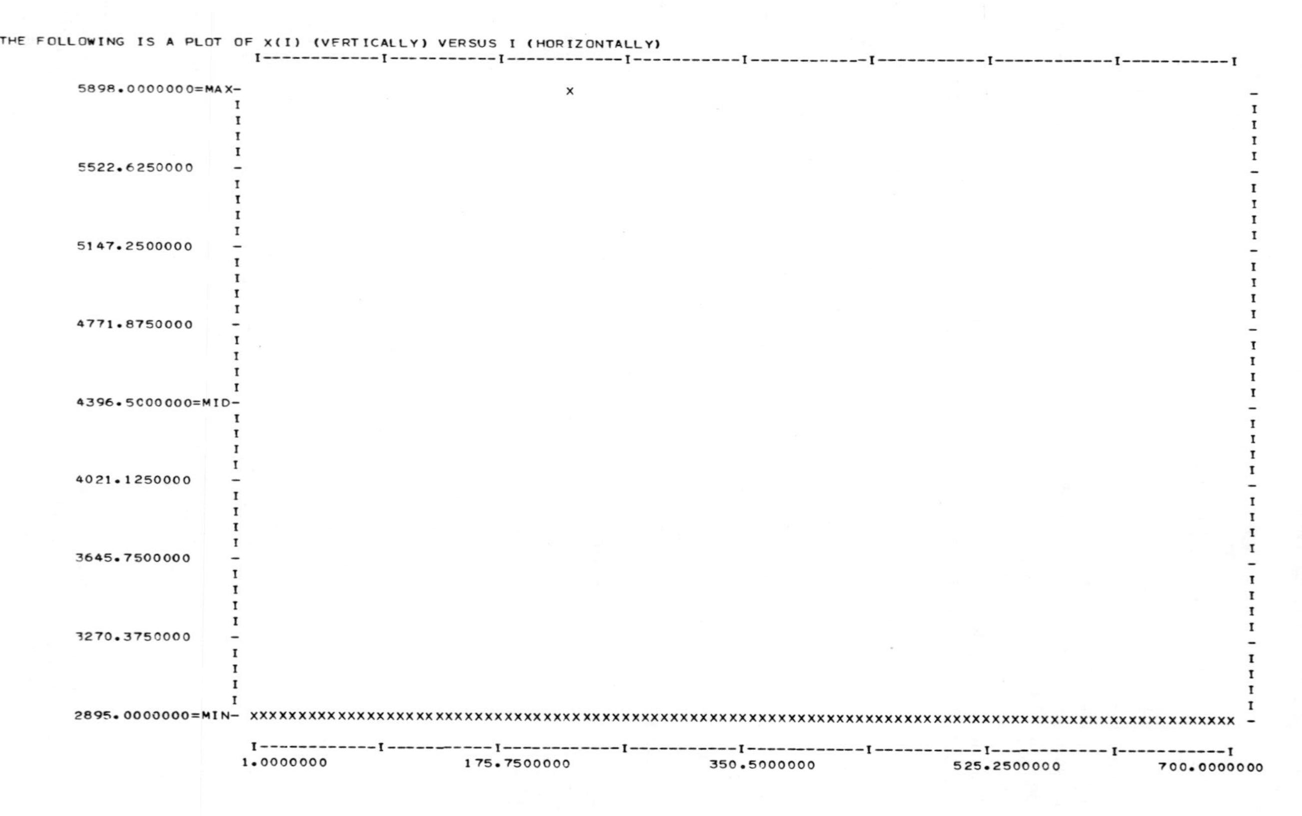

Figure 1a. Run sequence plot. Voltage counts data.

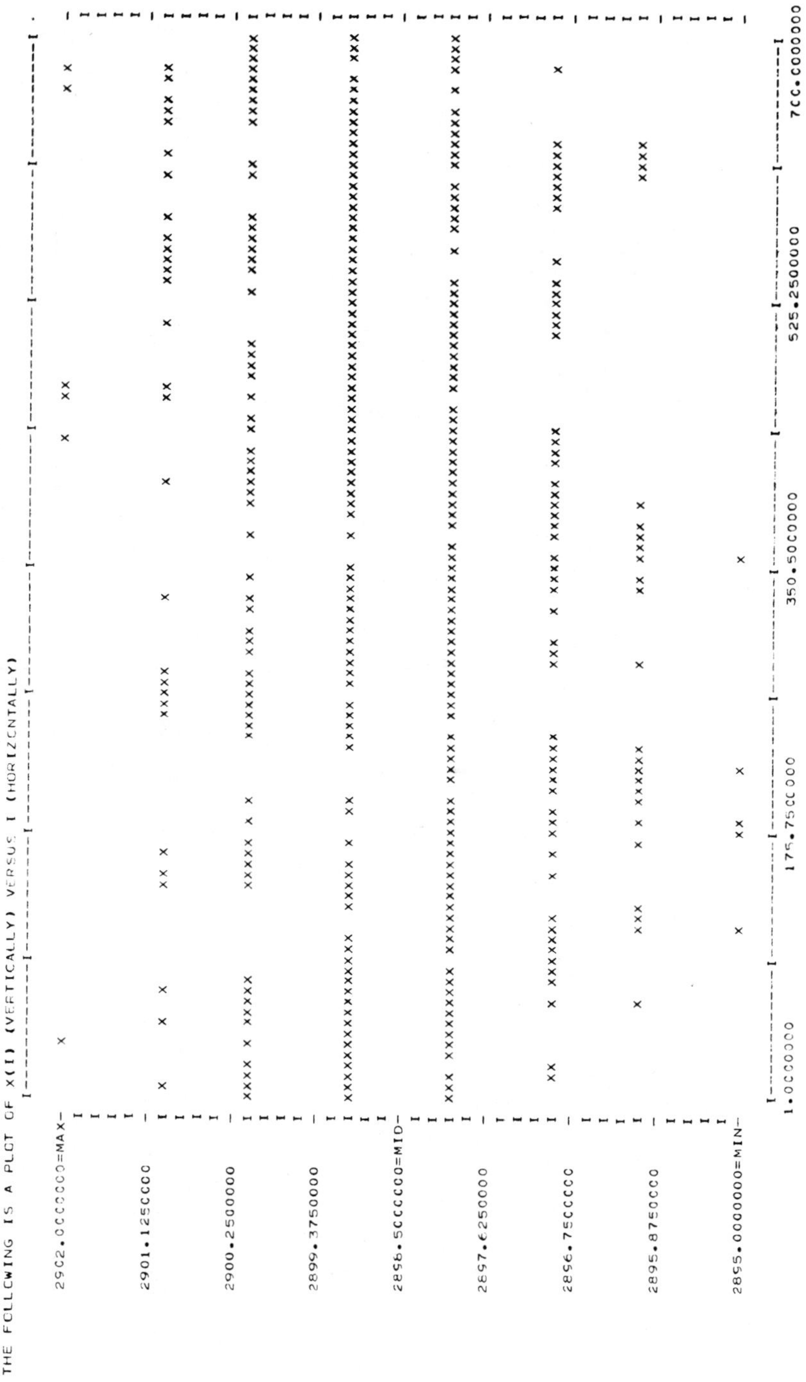

THE FOLLOWING IS A PLOT OF X(I) (VERTICALLY) VERSUS I (HORIZONTALLY)

2902.0000000=MAX
2901.1250000
2900.2500000
2899.3750000
2898.5000000=MID
2897.6250000
2896.7500000
2895.8750000
2895.0000000=MIN

1.0000000 175.7500000 350.5000000 525.2500000 700.0000000

Figure 1b. Run sequence plot. 1a. with outlier deleted.

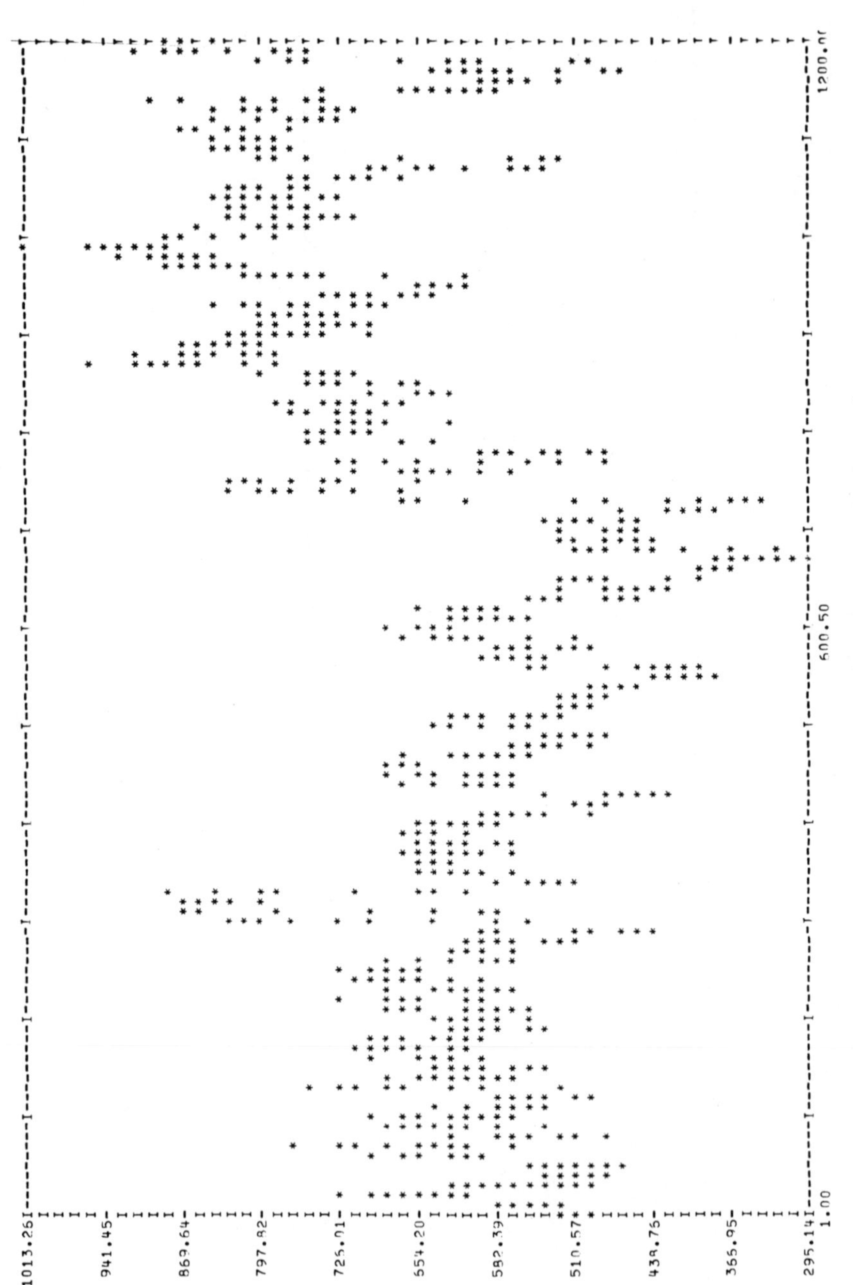

Figure 1c. Run sequence plot. Wind velocity data.

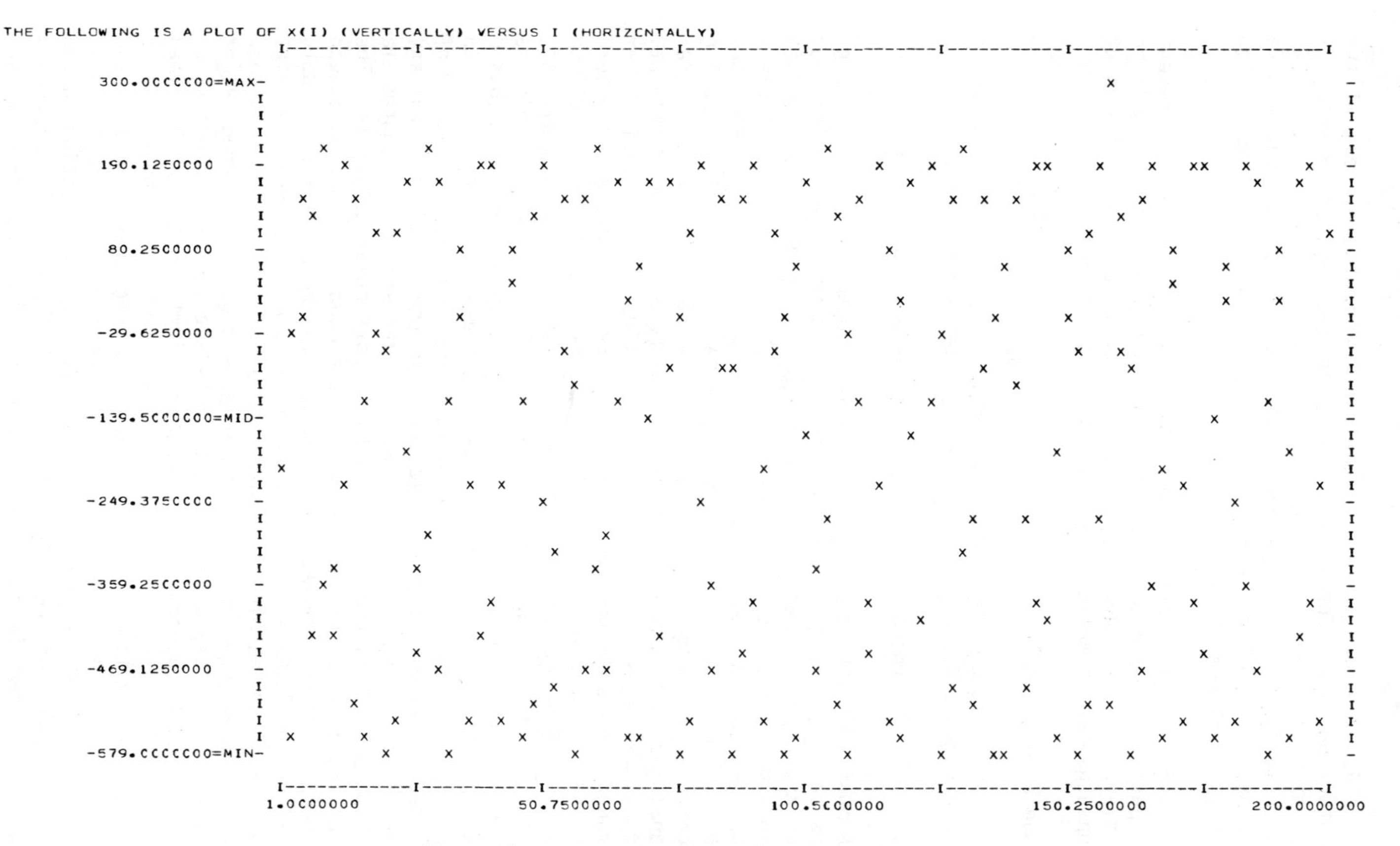

Figure 1d. Run sequence plot. Beam deflection data.

4. A plot allows the analyst to see and use all of the data--thus no information is lost in the typical exercise of forming statistics which (in essence) summarize and map information latent in the entire data set into a single number--a number which will usually only be sensitive to one particular analysis aspect of the data.

5. A plot allows the analyst to check many different aspects of the data simultaneously--and so information will be relayed not only about what is being investigated, but also about unsuspected anomalies (e.g., outliers) in the data.

RUN SEQUENCE PLOT

We assume that the experimentalist has collected n observations $Y_1, Y_2, \ldots, Y_n$ under the univariate model:

$$\text{response } Y_i = \text{constant } c + \text{error } e_i \quad (1)$$

Almost all data have a "time" run sequence (i=1, 2, ..., n) associated with it. Although the collection of data points may or may not have been equispaced in time, the ordering of the data in time (i.e., the run sequence) is usually well-defined (unless observations are simultaneously collected). In cases where the data acquisition rate is such that there is an equal time-spacing between collected data points, the run sequence has a natural analogue to a possibly relevant factor (time) in the experiment; in other cases, when the data acquisition rate is variable or random, no such analogue exists--yet the run sequence "factor" is still frequently of interest. The run sequence plot (defined as a plot of Y_i versus i) is the simplest possible data plot and yet is almost invariably informative. This run sequence plot is the recommended first step in assessing whether the basic assumptions of the measurement process are tenable. In particular, this plot yields information about the assumption of fixed location, fixed variation and the implicit corollary assumptions that the data set is outlier-free. Figures 1a through 1d illustrate this on three particular data sets. The two upper plots present data (voltage counts) from a Josephson Junction cryothermometry experiment. The upper left plot (data all on the lower margin with the exception of an isolated point on the upper margin) is indicative of an outlier (in this case due to a keypunch error in the leading digit). The upper right plot is the corrected data set; note the absence of shifts or variational changes as one proceeds left to right (in time) across the plot. (Note also how this plot gives the analyst a clear and initial "feel" for the discrete character of these data.) The lower left figure is a run sequence plot for wind velocity data. Note the apparent shift (up) in location in the second half of the data--a clear indication of a process apparently not in statistical control. The lower right figure gives a run sequence plot for deflections of a steel-concrete beam when a periodic force is applied to it. No apparent shift in location or variation and no apparent outliers are evident from this plot, and so this particular data set "passes" the scrutiny of this first statistical technique.

It is to be noted in passing that the FORTRAN calling sequence in the upper left corner of this figure (and most other figures in this paper) refer to calls to subroutines in the DATAPAC (6,7) data analysis package which produced the computerized output that comprises the figures.

An important generalization of the run sequence plot is the control chart. Rather than plotting x_i vs i (as above), the sample mean control chart plots $\bar{x}_i$ vs i when some arbitrary (but fixed) numbers of observations in the original sequence have been grouped together to form each $\bar{x}_i$. If the original process is normal, or if the number of observations grouped to form a single mean is large, then normal probability limits can be inserted onto the chart so as to define a typical band of variation for the process. Large (or frequent) excursions outside the band is an indication of a process that is no longer "in control." In addition to $\bar{x}$ control charts, other useful control charts are s (standard deviation) charts, r (range) charts and CUSUM (cummulative sum) charts. Collectively, they are excellent diagnostic tools for determining whether in particular an outlier has occurred and in general whether a shift in location or variation has occurred in the process. For further information on control charts, the reader is referred to Himmelblau (8).

LAG-1 AUTOCORRELATION PLOT

The lag-1 autocorrelation plot is defined as a plot of Y versus Y_{i-1} over the entire data set; that is, the following n-1 points are plotted: (Y_2, Y_1), (Y_3, Y_2), (Y_4, Y_3), ... (Y_n, Y_{n-1}). The lag-1 autocorrelation plot is sensitive to the randomness assumption in a measurement process. If the data are random, then adjacent observations will be uncorrelated and the plot of Y_i versus Y_{i-1} will appear as a data cloud with no apparent structure. However, if the data are not random and if adjacent observations do have some autocorrelation this structure will frequently manifest itself in the autocorrelation plot. Figure 2 gives three examples of the autocorrelation plot. The upper left plot is an autocorrelation plot of the aforementioned voltage counts--no apparent structure is noticeable (aside from the lattice effect due to the discreteness of the data). The upper right plot is for the wind velocity data--note the pronounced linear structure in this plot which implies that the randomness assumption is untenable for these data. The lower plot is for the beam deflection data--note the well-defined

```
THE FOLLOWING IS A PLOT OF X(I) (VERTICALLY) VERSUS X(I-1) (HORIZONTALLY)
                     I-------I-------I-------I-------I-------I-------I-------I-------I
2902.0000000=MAX-                                    X       X       X                -
                 I                                                                    I
                 I                                                                    I
                 I                                                                    I
                 I                                                                    I
2901.1250000    -                                                                     -
                 I                           X       X       X       X       X        I
                 I                                                                    I
                 I                                                                    I
                 I                                                                    I
2900.2500000    -                                                                     -
                 I                   X       X       X       X       X       X        I
                 I                                                                    I
                 I                                                                    I
                 I                                                                    I
2899.3750000    -                                                                     -
                 I                                                                    I
                 I                   X       X       X       X       X       X      X I
                 I                                                                    I
                 I                                                                    I
2898.5000000=MID-                                                                     -
                 I                                                                    I
                 I                                                                    I
                 I X                 X       X       X       X       X       X      X I
                 I                                                                    I
2897.6250000    -                                                                     -
                 I                                                                    I
                 I                                                                    I
                 I                                                                    I
                 I X                 X       X       X       X       X       X      X I
2896.7500000    -                                                                     -
                 I                                                                    I
                 I                                                                    I
                 I                                                                    I
                 I                   X       X       X       X       X                I
2895.8750000    -                                                                     -
                 I                                                                    I
                 I                                                                    I
                 I                                                                    I
                 I                                                                    I
2895.0000000=MIN-                    X       X       X                                -
                     I-------I-------I-------I-------I-------I-------I-------I-------I
                 2895.0000000      2896.7500000     2898.5000000     2900.2500000     2902.0000000
```

Figure 2a. *Lag-1 (Y_i vs. Y_{i-1}) autocorrelation plot. Voltage counts.*

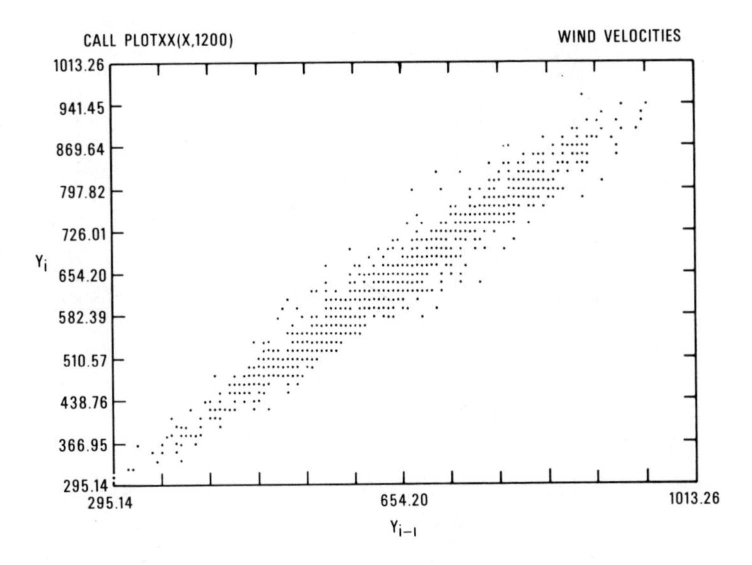

Figure 2b. Lag-1 (Y_i vs. Y_{i-1}) autocorrelation plot. Wind velocities.

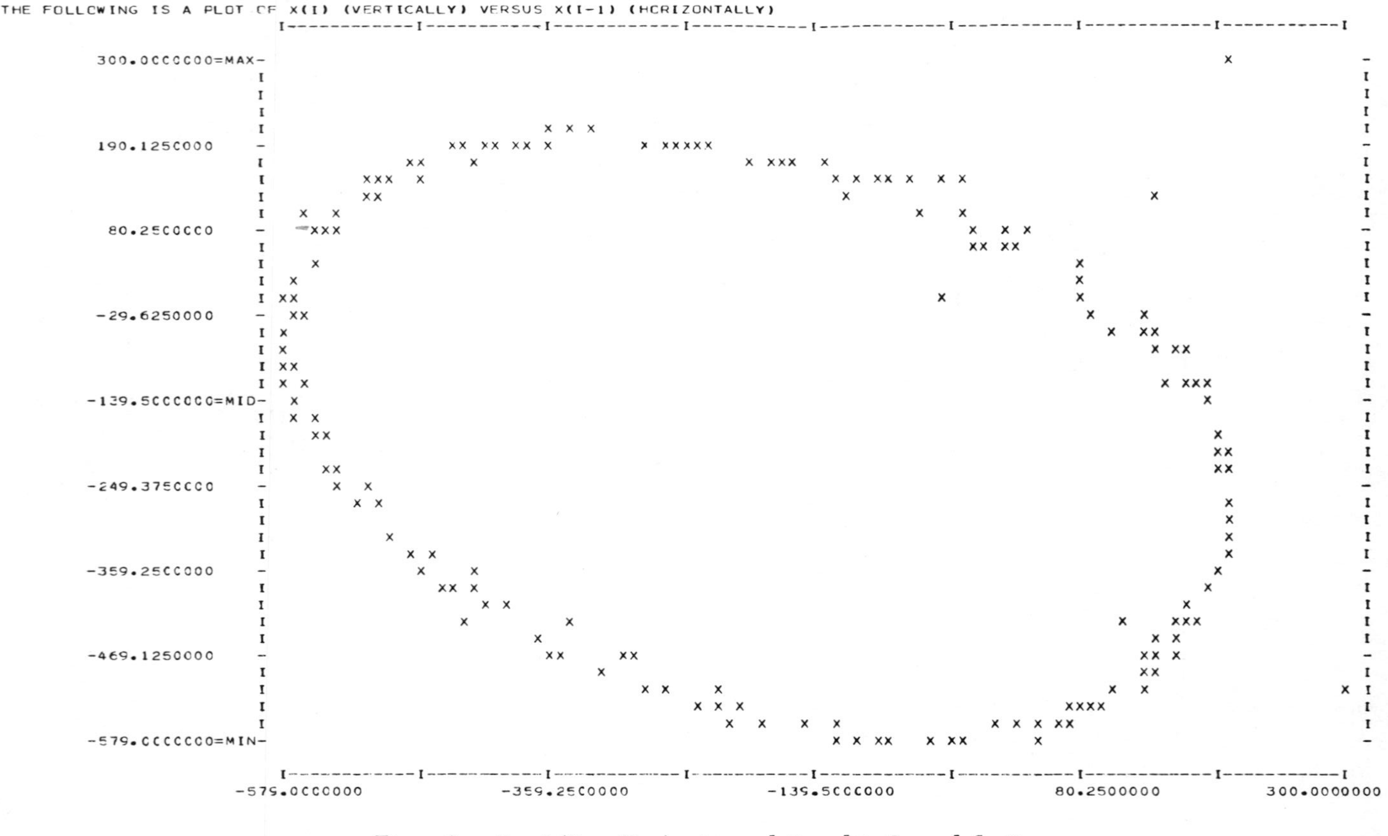

Figure 2c. *Lag-1 (Y_i vs. Y_{i-1}) autocorrelation plot. Beam deflections.*

elliptical structure of the autocorrelation plot which is also indicative of the untenableness of the randomness assumption. In this last case the lack of randomness was, as it turned out, due to an underlying cyclic structure in the data (*i.e.*, the true model was $Y_i = c + a \cdot \sin(\delta i + \phi) + e_i$ (where i is time) rather than the assumed $Y = c + e$. The reader should note the two points in the upper right portion of the plot which are off the ellipse. This is due to a single outlier in the data and demonstrates the secondary sensitivity of the lag-1 autocorrelation plot to outliers.

RUNS TEST

The runs test is a technique that is specifically used for testing randomness. The assumed underlying model for application of this technique is expressed by eq. (1). To illustrate the technique, consider the run sequence plot of 50 spectrophotometric transmittance data points in figure 3. It is apparent from the plot that the data are not random (note how observations 35 to 45 are not random but rather near-monotonic in nature). To scrutinize the correlation structure in this data set, consider the runs analysis given in figure 4. A run up of length i means that there are exactly (i+1) successive observations such that each observation is greater than (or at least equal to) the previous observation. The underlying theory behind the runs test is that if the data are random and *if* the sample size is known (in this case, n=50), the *number* of runs up of length 1, of length 2, *etc.*, may be considered as random variables whose expected values and standard deviations can be calculated from theoretical considerations (9) and these calculations will *not* depend on the (unknown) distribution of the data but only on its assumed randomness. Having computed such theoretical values, the final step in the test is to compute from the data the observed number of runs (up) of length 1, of length 2, *etc.*, and then determine how many *theoretical* standard deviations that this *observed* statistic falls from the theoretically expected value. This is most easily done by formation of the standardized variable:

$$\frac{N_i - E(N_i)}{SD(N_i)}$$

where N_i is the observed number of runs (up) of length i, $E(N_i)$ is the theoretical expected number of runs up of length i and $SD(N_i)$ is the theoretical standard deviation of the number of runs up of length i. This standardized variate is given in the right-most column of figure 4. For random data, one would expect values of, say, ±1, ±2, ±3 in this column, *i.e.*, the *observed* number of runs of length i should be only a few (at most) standard deviations away from the theoretical expected value for the number of runs of length i. For nonrandom data, the

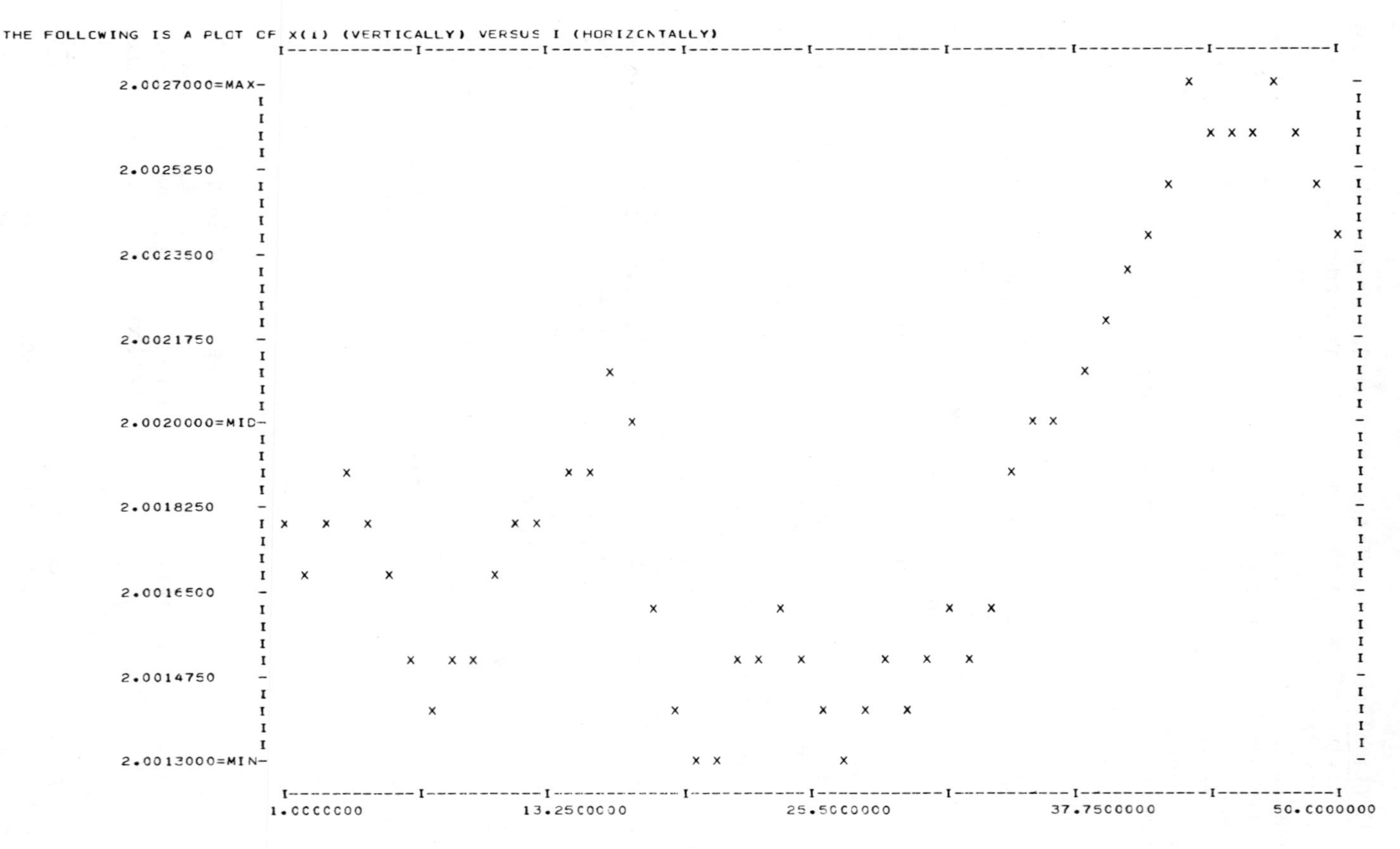

Figure 3. Run sequence plot for spectrophotometric measurement of transmittance

RUNS UP

STATISTIC = NUMBER OF RUNS UP OF LENGTH EXACTLY I

I = LENGTH OF RUN	VALUE OF STAT	EXP(STAT)	SD(STAT)	(STAT-EXP(STAT))/SD(STAT)
1	1.0	10.4583	3.2170	-2.94
2	3.0	4.4667	1.6539	-.89
3	1.0	1.2542	.9997	-.25
4	.0	.2671	.5003	-.53
5	.0	.0461	.2132	-.22
6	.0	.0067	.0818	-.08
7	.0	.0008	.0291	-.03
8	1.0	.0001	.0097	103.06
9	.0	.0000	.0031	-.00
10	1.0	.0000	.0009	1087.63
11	.0	.0000	.0003	-.00
12	.0	.0000	.0001	-.00
13	.0	.0000	.0000	-.00
14	.0	.0000	.0000	-.00
15	.0	.0000	.0000	-.00

STATISTIC = NUMBER OF RUNS UP OF LENGTH I OR MORE

I = LENGTH OF RUN	VALUE OF STAT	EXP(STAT)	SD(STAT)	(STAT-EXP(STAT))/SD(STAT)
1	7.0	16.5000	2.0696	-4.59
2	6.0	6.0417	1.3962	-.03
3	3.0	1.5750	1.0622	1.34
4	2.0	.3208	.5433	3.09
5	2.0	.0538	.2299	8.47
6	2.0	.0077	.0874	22.79
7	2.0	.0010	.0308	64.85
8	2.0	.0001	.0102	195.70
9	1.0	.0000	.0032	311.64
10	1.0	.0000	.0010	1042.19
11	.0	.0000	.0003	-.00
12	.0	.0000	.0001	-.00
13	.0	.0000	.0000	-.00
14	.0	.0000	.0000	-.00
15	.0	.0000	.0000	-.00

Figure 4. Runs analysis for spectrophotometric measurement of transmittance. Call runs. ($\times 50$)

deviations from the expected values will, of course, be much larger and this is the crux of the runs test. Note that in the spectrophotometric data, the randomness assumption is entirely untenable as indicated by the excessively large values of the standardized statistic in the last column (*e.g.*, the number of runs up of length 10 in the data is 1 and yet for n=50 observations and for random data, we should have essentially no runs up of length 10--this 1 run up of length 10 is over 1000 standard deviations from its expected value and so the randomness assumption must be rejected).

The above-described runs analysis is a valuable additional tool for testing the specific hypothesis of randomness.

The net effect of the above is that if one were to report the sample mean X of the 50 observations as the "final result" of this experiment, the uncertainty statement associated with X (for example $s_{\bar{X}}$ = the estimated standard deviation of $\bar{X}$) would certainly have to be based on many fewer degrees of freedom than n-1 = 49. As is evident from the data, there are not 50 *independent* observations of the transmittance; there would be considerably fewer--and this would result (from a practical point of view) in a larger (and more realistic) value for $s_{\bar{X}}$.

BAND PLOTS

The assumed underlying model for application of this technique is again as in eq. (1). To graphically test this model, however, an alternative model, *viz.*,

$$\text{response } Y_i = f(X_{1i}) + \text{error } e_i \qquad (2)$$

where $f(X_{1i})$ is some unknown function of the variable X_1, and where X_1, is a possible variable affecting the response. A band plot *(4)* is a specially-constructed plot of the response variable Y versus another variable X . A band plot considers all the data within various classes of the horizontal axis variable and then, rather than plotting all such points, summarizes each subset of data into five statistics: the median, the lower and upper quartiles, and the two extremes (minimum and maximum). A line connecting the medians across the horizontal axis adds continuity to the plot and gives a more robust indication of whether the response variable shifts location with respect to the horizontal axis variable. Lines connecting the various lower quartiles provide a lower practical limit to the "body" of the data whereas lines connecting the upper quartiles delineate an upper edge to the body of the data. The *flatness* (lack of trend) of the band between upper and lower quartiles is an indication of whether or not model 1 above (the fixed location model) is tenable. The *width* of the band between upper and lower quartiles is an indication of whether the fixed variation assumption (with respect to the horizontal axis variable) is tenable.

The example that will be used for the band plot will demonstrate how it can (in certain special circumstances) be used to test randomness. The data set consisted of 400 percentage measurements taken from a near-complete surface inspection of a circular austenite standard reference material specimen. A given reading is the percentage austenite value for that particular small sub-area of the specimen. To test the hypothesis that the specimen was homogeneous (that is, that 2-dimensional randomness existed), a band plot of the percentage austenite readings versus angle (from some reference radial spoke of the circular specimen) was constructed. Figure 5 illustrates the resulting band plots when this angle factor was divided into 24 classes with a class width of 15 degrees was used. Thus, all data in a given 15 degree wedge were assembled and then summarized into five statistics: median, lower quartile, upper quartile, minimum, and maximum; these five statistics were then plotted to represent that specimen wedge rather than plotting all the data in the wedge. If the sample were homogeneous (with respect to angle), the band plot should be near-flat over the entire 360° range. As the plot illustrates, this is not the case for these data--the percentage austenite measurements tend to be low in the vicinity of 135°, tend to be high near 280°, and tend again to be low near 330°. The plot clearly shows the (homogeneity) randomness assumptions to be suspect for this specimen.

2-VARIABLE GRAPHICAL ANALYSIS OF VARIANCE

This technique is applicable when a multi-factor model of the following type is suspected:

response Y_{ij} = constant c + $B_{1i}X_{1i}$ + $B_{2j}X_{2j}$ + error e_{ij}

with an alternative general model of the following form:

$$\text{response } Y_{ij} = f(X_{1i}, X_{2j}) + \text{error } e_{ij}$$

where f is an <u>unknown</u> function relating the nonrandom variables X_1 and X_2 to the response variable Y, and where X_{1i} and X_{2j} indicate different discrete limits within the variables X_1 and X_2, respectively. Although the plot of Y versus X_1 will certainly give the analyst some indication of the nature of the function f, the main point is whether and how the response is additionally affected by the second variable (say variable X_2) for some (but not all) values of the variable X_1. If the variable X_2 affects the response for only some (but not all) of the values of the variable X_1, in statistical terms this is referred to as an "interaction" existing between X_1 and X_2--thus the effect of X_2 on the response is dependent on the value of X_1. Rather than apply the usual 2-factor analysis of variance

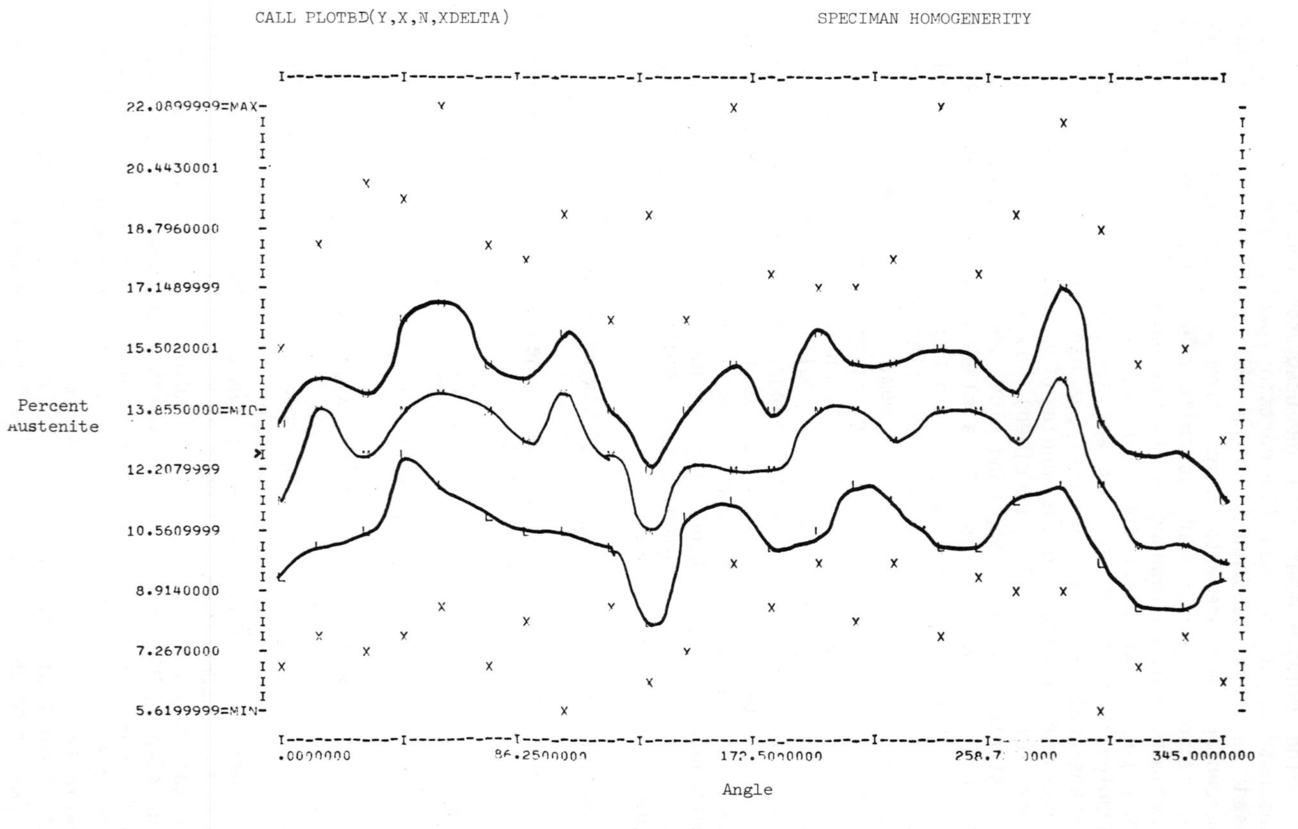

Figure 5. Band plot for specimen homogeneity data

(ANOVA) to data from this model, we apply the graphical procedure illustrated in figure 6. This graphical analysis of variance (GANOVA) (10,11) is simply a plot of the response variable versus one factor, with different levels of the second factor indicated by different types of plot characters within the plot. The GANOVA procedure is very revealing in that it communicates all of the latent relevant information in this 2-factor system.

This technique is illustrated in figure 6 which plots the residuals from a fit of the response variable (days to failure) from a stress fatigue experiment versus lab (9 labs) with the value of the plot character representing various levels (3 levels) of a second variable (experiment configuration) which could (but hypothetically should not) affect the response. A plot character value of 3, *e.g.*, indicates that this particular data point was generated from configuration 3 of the experiment. The two independent variables are:

laboratory (9 levels--plotted horizontally).
configuration (3 levels--denoted by different plot characters).

Making reference to figure 6, it is seen that the assumption that all levels of the configuration factor affects the response in a uniform fashion is untenable. It is clear from the plot that a lab-configuration interaction exists. For example, configurations 2 and 3 yield consistently low values for lab 4 while configuration 1 yields low and rather variable values for labs 5 and 7. A suspicious low observation also is seen to exist for lab 3, configuration 2. Such an augmented plot--as described above--is a useful technique for examining the assumption that the response is *not* dependent on some particular variable.

It is to be noted in passing that although the plot character is the recommended procedure for conveying information about the second variable, one could also just as well use the type of line for conveying the second-variable information. The former is recommended when generating computer printer plots--which are by nature *discrete*. The latter is recommended when a *continuous* printing device (*i.e.*, one capable of drawing different types of *continuous* lines) is available.

Figure 7 illustrates the above line-type alternative with an example based on measured voltages from electrical connectors. The two independent variables here are:

time in days (plotted horizontally)
connector type (3 levels--denoted by different line types)

The multiple lines of the same type are due to experimental replication. Although much can be said about the plot and about

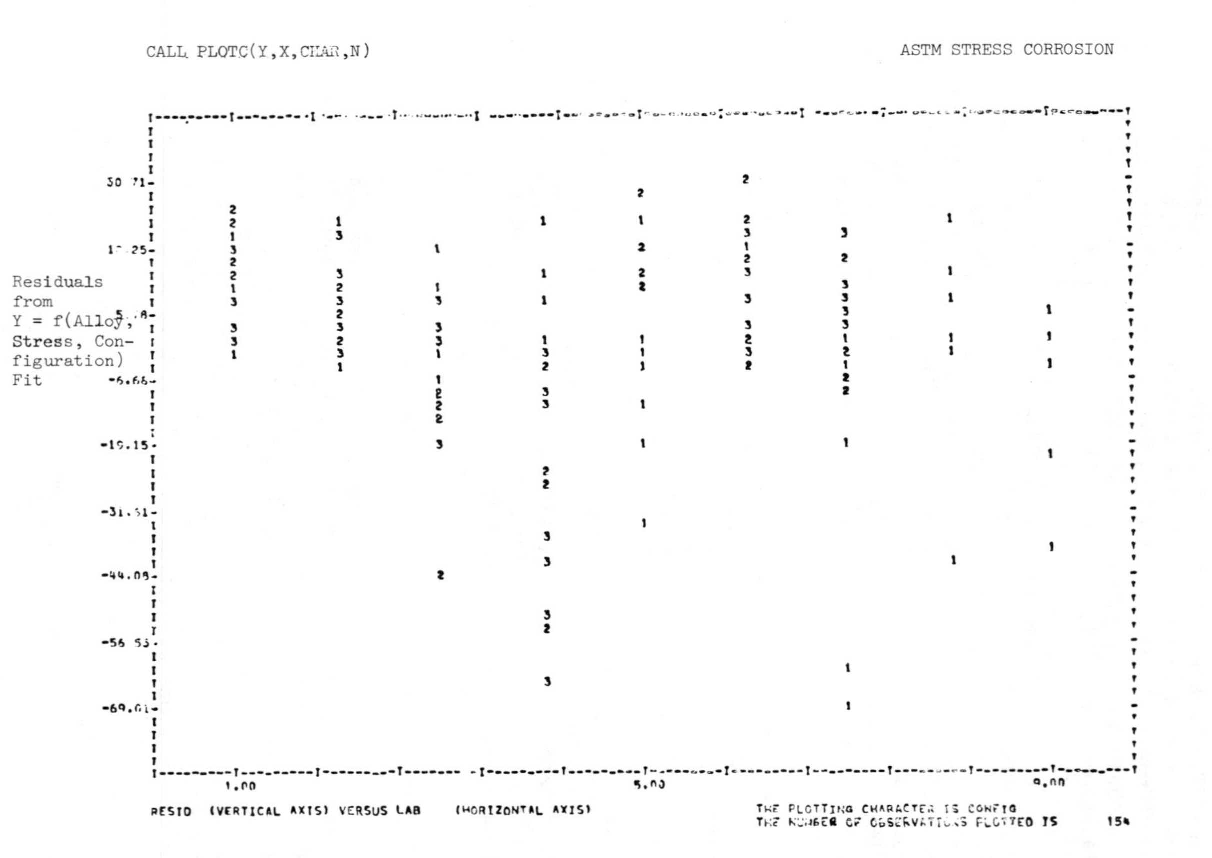

Figure 6. Two graphic analyses of variance for ASTM stress corrosion data

the relative effects of the two factors on the response, we concentrate herein on violations of basic assumptions and note that an apparent violation in the form of an outlier is evident from the plot--note how the fourth data point of the bottom line of the plot is inconsistent with the other data lines in this bottom group. This fourth point is clearly an outlier, and yet its detection may very easily have been lost in the numerical mechanics of a standard ANOVA.

3-VARIABLE GANOVA

This 3-variable GANOVA technique is applicable where a multifactor model is appropriate, i.e., the underlying hypothesized model is of the form (e.g., for three factors):

$$\text{response } Y_{ijk} = \text{constant } c + B_{1i}X_{1i} + B_{2j}X_{2j} + B_{3k}X_{3k} + \text{error } e_{ijk}$$

with an alternative general model of the following:

$$Y_{ijk} = f(X_{1i}, X_{2j}, X_{3k}) + \text{error } e_{ijk}$$

with f unknown, where the doubly-subscripted B's refer to factor effects and the doubly-subscripted X's refer to coded dummy levels of each factor. Again, rather than apply the standard 3-factor analysis of variance (ANOVA) to data from this model, we apply the graphical procedure illustrated in figure 7. This graphical analysis of variance (GANOVA) (10,11) is a generalization of the type of plot discussed in section 6 and is defined simply as a plot of the response variable versus one factor, with different levels of the second factor indicated by different types of plot characters within the plot, and with the different levels of the third factor indicated by different types of lines within the plot. The 3-variable GANOVA conveniently communicates at a glance all of the relevant information in this 3-factor system.

Figure 8 illustrates the application of the 3-variable GANOVA to drill thrust force. These data are drawn from the excellent article by Hamaker (10) whose main point was exactly as we are emphasizing here--viz., that there exist valuable graphical alternatives to the usual ANOVA. The three independent variables are:

drill speed (5 levels--plotted horizontally)
material (2 levels--denoted by different plot characters)
feed rate (3 levels--denoted by different line types)

Referring to figure 8, one concludes that variables 2 (material) and 3 (feed rate) both affect the response in a non-negligible

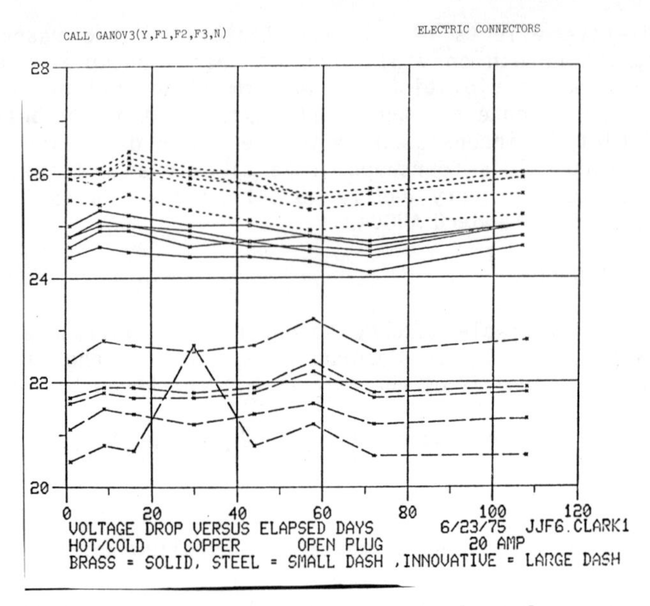

Figure 7. Three graphic analyses of variance for electrical connectors data

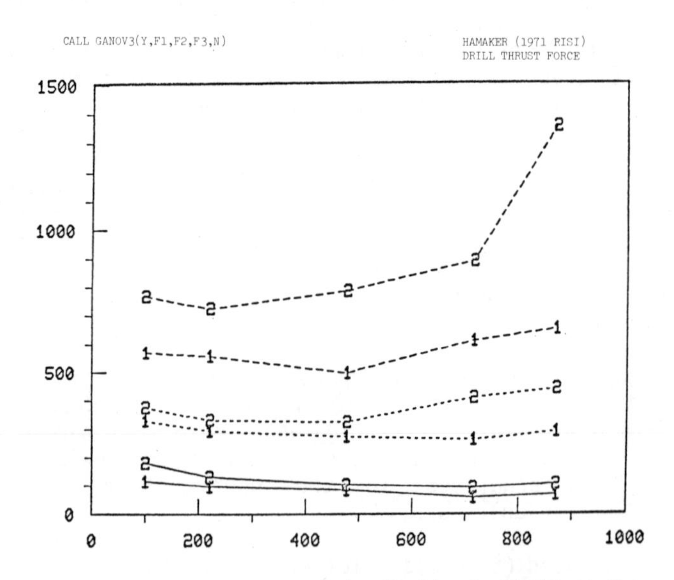

Review of the International Statistical Institute

Figure 8. Three graphic analyses of variance for Hamaker (10). Drill thrust force data; thrust force (v) vs. drill speed (x); plot character = material (2 levels); type of line = feed rate (3 levels).

way. The apparent existence of an outlier is also evident from the plot.

It is to be noted that due to the necessary use of line types, the 3-variable GANOVA can be done only with the continuous printing devices.

YOUDEN PLOT

The Youden plot (12,13) is a useful graphical technique most commonly applicable to interlaboratory experiments when there exists exactly 2 runs (or 2 specimens, or 2 levels of some particular factor, *etc.*) to be tested for a certain property of interest. Ideally, if the 2 runs are "alike" and if all laboratories are "alike," the response model should be as in eq. (1); whereas if a laboratory effect existed, an appropriate model might be:

$$\text{response } Y_i = \text{constant } c + L_i + \text{error } e_i$$

(where L_i represents an effect due to laboratory i--i = 1, 2,...,k) and additionally, if both a laboratory effect and a run effect existed, an appropriate model might be:

$$\text{response } Y_i = \text{constant } c + L_i + R_j + \text{error } e_{ij}$$

(where R_j represents a run effect for run j--j = 1, 2).

To test which model is appropriate, a Youden plot is applied which is defined as a plot of the k (where k = the number of laboratories in the experiment) coordinate pairs: Y_{11}, Y_{12}), (Y_{21}, Y_{22}),...(Y_{k1},Y_{k2}), where Y_{ij} represents the measured values obtained from laboratory i (i = 1, 2, ..., k) on run j (j = 1, 2).

To facilitate the graphical analysis, the plot character is again used to "pack" in extra information--in this case, about the laboratory factor. Thus, *e.g.*, a plot character of 4 indicates that the measurement in question came from laboratory 4.

The Youden plot is illustrated in figure 9 as applied to data from an ASTM stress corrosion experiment where 7 (k) laboratories were being tested.

If no laboratory or run effects existed, the resulting Youden plot will appear as a random 2-dimensional scatter of points. Alternatively, if laboratory and/or run effects do exist, much useful information about the nature of such effects can be gleaned from the resulting plot. The plot in figure 9 actually is based on 7 x 5 = 35 plot points (not all of which

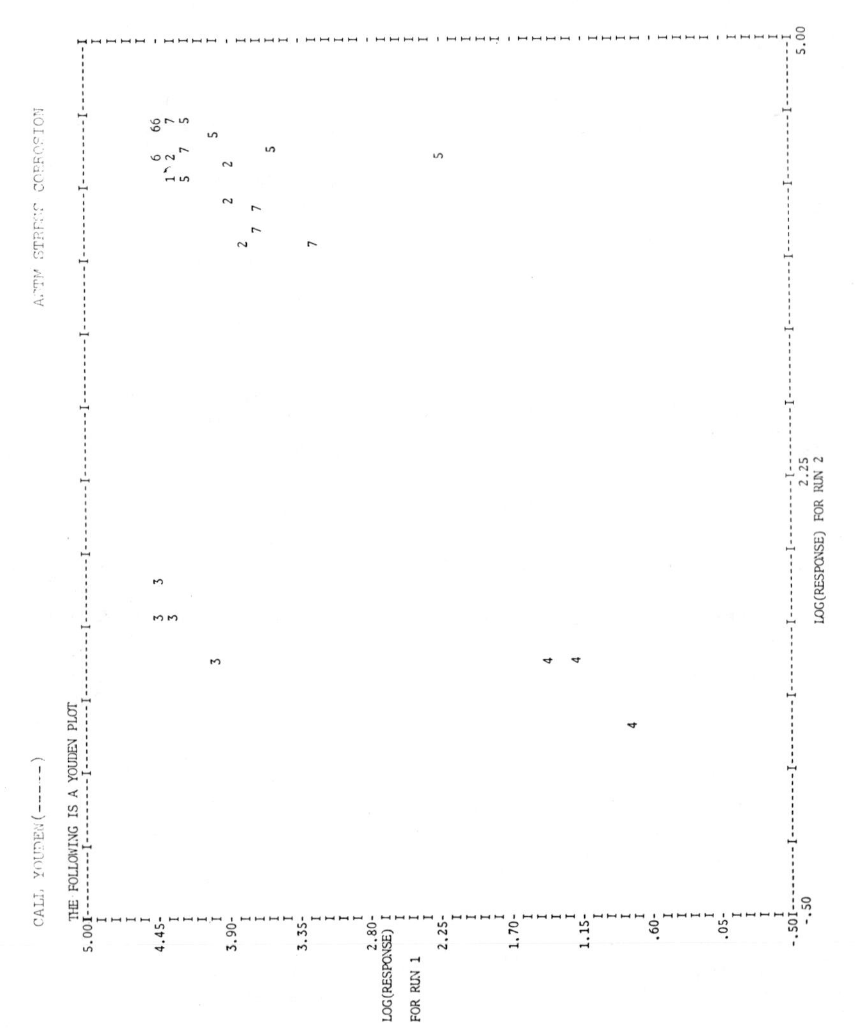

Figure 9. Youden plot; plot character = laboratory

appear due to computer printer overstriking). The multiplicity of 5 is due to the existence of 5 replications per lab--such replications pose no problems in utilizing the Youden plot.

With respect to how to interpret a Youden plot, several characteristics are to be noted. A displacement of points from the same laboratory along the 45° diagonal is indicative that this laboratory is consistently generating low (or high) readings relative to the other laboratories (the cluster of laboratory 4 points in figure 9 is illustrative of this negative laboratory bias). On the other hand, a cluster of points from the same laboratory displaced off the diagonal represents inconsistent readings by that laboratory from one run to the next. Figure 9 indicates that laboratory 3 has such an inter-run variability problem. For laboratory 3, the readings for run 1 are consistently higher than those for run 2.

The Youden plot is a simple--yet extremely effective--method for analyzing interlaboratory data.

EXAMINING DISTRIBUTIONAL INFORMATION

The discussion has already touched on three (randomness, fixed location, fixed variation) of the four assumptions typically made about a measurement process. The fourth assumption (fixed distribution) will now be addressed.

From a statistical point of view, there are five reasons why distributional information should be routinely checked:

1. optimal estimators for location and variation parameters;

2. validity of critical values used in statistical tests of significance;

3. assessment of goodness of fit in regression;

4. existence of outliers;

5. assessment of whether the measurement process is in control.

The last of these reasons (assessment of whether a measurement process is in statistical control) is the main one with respect to the overall purpose of this paper. The first four reasons provide additional motivation for checking distributional assumptions, and will be individually touched on at this time.

The first (optimal estimators) point refers to the case where one is interested in estimating from a given data set the location parameter c and variation (dispersion or scale) parameter σ in the model described in eq. (1). (It is assumed that the error, e_i is a random variable with mean 0 and (unknown) standard deviation, σ.) Various estimators of c would, for example, include the usual sample mean of n observations $\hat{c} = SY_i/n$, the sample median ($\hat{c}$ = the middle observation in the ordered set of observations), or the sample midrange ($\hat{c}$ = the average of the smallest and largest observations). It is a statistical "fact-of-life" that in estimating location and variation parameters, the goodness (accuracy) of a particular estimator and the choice of an optimal estimator are dependent on the underlying distribution. For example, if the underlying distribution which generated the data set were the bell-shaped (normal or Gaussian), the best estimator of c would be the sample mean. However, if the underlying distribution were uniform (i.e., it had a flat--rather than bell-shaped probability function), it can be theoretically demonstrated that the sample midrange, $\hat{c}$ = (smallest + largest)/2 is a much more accurate estimator of c than the sample mean. Alternatively, if the underlying distribution for the data were, e.g., very "long-tailed" like the Cauchy (i.e., the probability function is bell-shaped but higher valued in the tails than the normal), then theory dictates and practice confirms that the sample median is a much more accurate estimator of c than either the sample mean or the simple midrange. Thus, it is seen that for estimating the constant c in the simplest possible response model (Y = c + e), a necessary preliminary step is to "estimate" the underlying distribution. Although the central limit theorem provides a theoretical basis for suggesting that for many physical science experiments, the normal distribution "should" be the underlying distribution, such normality should never be automatically assumed. As will be seen in the remaining sections, statistical techniques do exist which allow the analyst to easily and routinely check such distributional models.

The second reason why distributional information should be checked deals with the validity of test statistics. In the multifactor statistical techniques referred to as regression and analysis of variance, there are a variety of test statistics (mostly t and F statistics) which are applied to test the significance of various factors in the multi-factor model. It is an important statistical fact that the validity of these test statistics holds only if the residuals (deviations) after the fit are normally distributed. That is to say, it is the distributional characteristic of the residuals after the fit that dictate the validity of the t and F statistics. If the true underlying distribution of the residuals is non-normal, this will affect the true significance levels of the test statistics. The net result is that ultimately the conclusions about the significance of various factors in regression and ANOVA may be incorrect. Again, as emphasized before, no blind assumptions need be made about the distribution of such residuals. Techniques will be demonstrated to allow the distribution to be routinely checked.

The third reason for checking distributional information is related to the aforementioned regression and ANOVA. The point to be emphasized is that an additional important reason for examining the distribution of residuals after the fit is to determine whether or not one has arrived at a reasonable deterministic or functional model for the data. If the fitted regression or ANOVA model is correct, the residuals after the fit should ideally have the same four properties as has been previously discussed with respect to the univariate response variable, viz.:

random
fixed location
fixed variation
fixed distribution

In a large majority of cases, the residuals after the correct fit will not only follow some fixed distribution, but will also rather specifically follow a normal distribution. The implication of course is that in order to assess whether or not one has a correct fit, one ought to examine the distribution of the residuals to check for such normality. Though not a sufficient condition in itself for adequate fit, the normality of the residuals serves as a practical necessary condition which may profitably be used in determining model adequacy. From a pragmatic point of view, this third reason for examining distributional information is an extremely important one.

The fourth reason for checking distributional information deals with the outlier problem. How does one tell if a suspicious-looking observation is in fact an outlier? ("Outlier" as here used refers to an observation that was generated from a different model or a different distribution than was the main "body" of the data.) Frequently, an outlier will manifest itself in one or another of the plots already discussed in previous sections. However, an additional and at times more sensitive check is given by a detailed examination of the distribution of the data. An observation which appears to be a borderline outlier in some previous plots frequently turns out to be a well-defined outlier when examined relative to the distribution of the rest of the data. The same numerical observation may very well be a "typical" extreme observation relative to one distribution but an outlier relative to another distribution. By examining the distribution of the data (and/or the residuals after a fit), the analyst gives himself a much more sensitive tool for outlier detection and identification.

The fifth and final point with respect to the importance of checking for distributional information deals with the main point of this paper--predictability and the determination of whether a process is "in control." Predictability means being able to make probability statements about future output from the process. These probability statements will most commonly refer to expected variation (about some typical value) of output from the process. The main point is that such probability statements will change depending on the true underlying distribution of the process. A statement such as:

> "97-1/2% of the future observations from this measurement process should fall within (approximately) 3 standard deviations of the mean"

will of course be true if the underlying generating distribution is normal but on the other hand will be false if the underlying distribution is (for example) uniform, Cauchy or exponential. It is important for analysts to keep in mind that for non-normal distributions, a probability statement about expected future occurrences (*e.g.*, within two standard deviations of the mean) will change from distribution to distribution. The exact probability value (≅ 97-1/2% for the normal) must be (and can be) determined once the underlying distribution is determined. It is a recurring requirement to "estimate" the underlying distribution.

With these motivations and justifications for examining distributional information, the next two sections will present various data analysis techniques to carry out such examinations.

PROBABILITY PLOTS

A probability plot (14,15,16,17,18,19,20,21) is a graphical tool for assessing the goodness of fit of some hypothesized distribution (*e.g.*, normal, uniform, Poisson, *etc.*) to an observed data set. In describing a probability plot, it will be assumed that the model is as indicated in eq. (1). However, it is to be kept in mind that the probability plot technique has much greater generality inasmuch as it can be applied to the residuals after any multifactor fit as well as to the raw observations from the simple $Y_i = c + e_i$ model.

A probability plot is (in general) simply a plot of the *observed ordered* (smallest to largest) *observations* Y_i on the vertical axis versus the corresponding *typical ordered observations* M_i based on whatever distribution is being hypothesized. Thus, for example, if one were forming a *normal* probability plot, the following n coordinate plot points would be formed: (Y_1, M_1), Y_2, $M_2)$,...(Y_n, M_n) where Y_1 is the observed smallest data point, and M_1 is the theoretical "expected" value of the smallest data point from a sample of size n normally distributed points. Similarly, Y_2 would be the second smallest observed value and M_2 would be the "expected value" of the second smallest observation in a sample of size n normally distributed points. This proceeds up to Y_n which would be the largest observed data value and M would be the "expected value" of the largest observation in a sample of size n from a normal distribution. Thus, in forming a normal probability plot, the vertical axis values depend only on the observed data, while the horizontal axis values are generated independently of the observed data and depend only on the theoretical distribution being tested or hypothesized (normality in this case) and also the value of the sample size n. A probability plot is thus in simplest terms a plot of the observed versus the theoretical or "expected."

The crux of the probability plot is that the ith ordered observation in a sample of size n from some distribution is itself a random variable which has a distribution unto itself. This distribution of the ith ordered observation can be theoretically derived and summarized (i.e., mapped into a single "typical value") as can any other random variable. One can then pose the relevant question as to what single number best typifies the distribution associated with a given ordered observation in a sample of size, n. A computational disadvantage to the use of the mean is that different integration techniques may be needed for different types of distribution. For some distributions the mathematical integration does not exist. These considerations dictate that the median is superior to the mean in terms of forming a theoretical "expected" or "typical" value to summarize the entire distribution of the ith ordered observation in a sample of size n from the distribution being tested. Thus, to be precise, the M_i on the horizontal axis of the probability plot is taken to be the median of the distribution of the ith ordered observation in a sample of size n from whatever underlying distribution is being tested.

It is to be noted that the set of M_i as a whole will change from one hypothesized distribution to another--and therein lies the distributional sensitivity of the probability plot technique. For example, if the hypothesized distribution is uniform, then a uniform probability plot would be formed and the M_i will be approximately equi-spaced to reflect the flat nature of the uniform probability density function. On the other hand, if the hypothesized distribution is normal, then the M_i will have a rather sparse spacing for the first few $(M_1, M_2, M_3,...)$ and last few $(..., M_{n-2}, M_{n-1}, M_n)$ values but will become more densely spaced as one proceeds toward the middle of the set (..., $M_{n-1/2}$, $M_{n/2}$, $M_{n+1/2}$,...). Such behavior for the M_i is of course reflecting the bell-shape of the normal probability density function.

In summary, for a specific hypothesized distribution, D_0 the ith value M_i in the corresponding probability plot is a theoretical (but computable) value close to what one typically would "expect" for the value of the ith order observation <u>if</u> in fact one had taken a random sample of size n from the distribution D_o.

How does one use and interpret probability plots? In light of the above, it is seen that if in fact the observed data do have a distribution that the analyst has hypothesized, then (except for an unimportant location and scale factor which can be determined after the fact) the Y_i and M_i will be <u>near-identical</u> for all i, that is, over the entire set. Consequently, the plot of Y_i versus M_i will be <u>near-linear</u>. This linearity is the dominant feature to be checked for in any probability plot. A linear probability plot indicates that the hypothesized distribution, D_0 gives a good distributional fit to the observed data set. This combination of simplicity of use along with distributional sensitivity makes the probability plot an extremely powerful tool for data analysis.

The next logical question to be examined is what will the probability plot look like if the hypothesized distribution, D_0 is not correct--<u>i.e.</u>, if the underlying distribution that generated the data is not the same as the distribution, D_0 hypothesized by the analyst. In this case, the Y_i and M_i will <u>not</u> match over the entire set and so the resulting probability plot will be <u>nonlinear</u>. A very useful aspect of the probability plot is that the type of nonlinearity exhibited by a given probability plot will give the analyst useful information as to how the distributional hypothesis, D_0 should be <u>adjusted</u> so as to arrive at a better distributional fit to the data. This last point is an important asset of the probability plot technique for testing assumptions in distribution. For example, if the analyst believes that the true underlying distribution is in general a symmetric distribution (<u>i.e</u>., a distribution which has a probability function as illustrated in fig. 10) as opposed to a skewed distribution (<u>e.g.</u>, with a probability function as illustrated in fig. 11), then the probability plot analysis to be presently described is rather typical. The first step in such analysis is usually to test the normal distribution hypothesis (the normal being the most commonly-employed symmetric distribution) by forming a normal probability plot. In forming such a plot, let us consider the following five types of most commonly-encountered appearances of the normal probability plot: linear, S-shaped, N-shaped, nonsymmetric cross-over, and convex (see fig. 12).

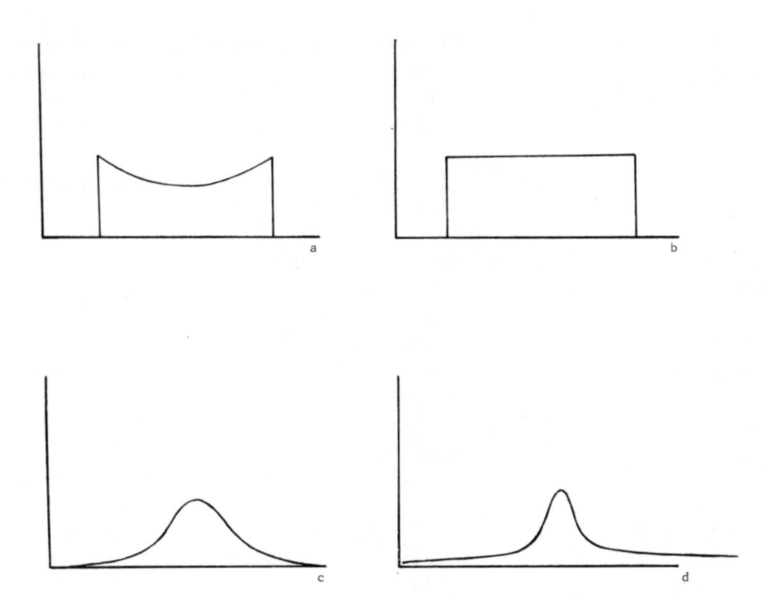

Figure 10. Symmetric probability density functions (pdf). (a.) U-shaped pdf; (b.) flat pdf; (c.) bell-shaped pdf; (d.) long-tailed-bell-shaped pdf.

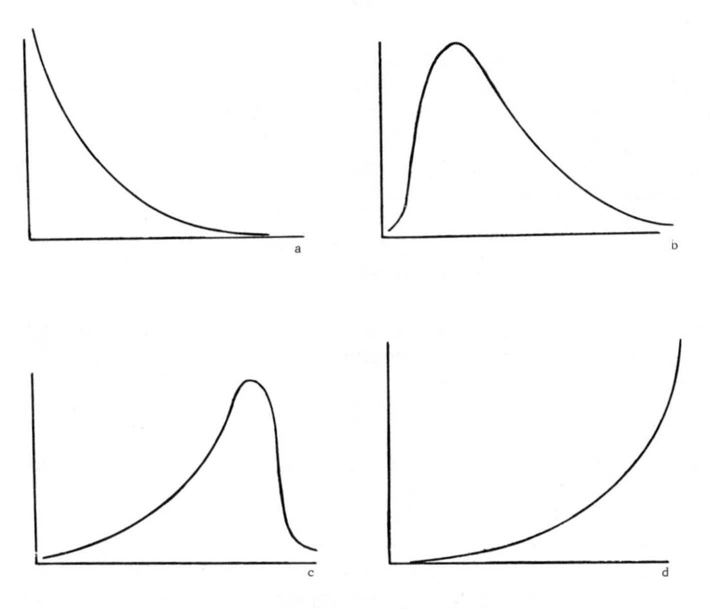

Figure 11. Skewed (i.e., asymmetric) probability density functions: a and b skewed to the left; c and d skewed to the right.

If the normal probability plot has the linear appearance of figure 12a, this indicates that the normal distribution yields an acceptably good fit to the data; so no further probability plots need be formed and the distribution analysis is completed.

If the normal probability plot has the S-shaped appearance of figure 12b, this indicates that the D_0 = normal hypothesis is incorrect, and that the true underlying distribution for the data is symmetric but is shorter-tailed than normal. Examples of such symmetric distributions, shorter-tailed than normal, would be a U-shaped distribution, a uniform distribution, or a truncated bell-shaped distribution. (These three distributions have probability functions as illustrated in fig. 13.) In such a case, the second iteration by the analyst would be to form an additional probability plot for some shorter-tailed distribution (e.g., from a uniform probability plot). If the resulting uniform probability plot is still S-shaped, the third iteration is to form a probability plot for a distribution that is even shorter-tailed than uniform (e.g., some U-shaped distribution). On the other hand, if the uniform probability plot has a form as in figure 12c (and which will be represented very crudely as an "N shape"), the third iteration would be to form a probability plot for some distribution shorter-tailed than normal but longer-tailed than uniform. Such iteration is continued until there is convergence to an acceptable linear probability plot. In practice, the analysis will usually converge to an acceptable distribution in a relatively small number of iterations.

To consider another possibility, if the original normal probability plot has the "N-shaped" appearance of figure 12c, this suggests that the D_0 = normal hypothesis is incorrect, and that the true underlying distribution for the data is still symmetric but is longer-tailed than normal. An example would be the Cauchy (also known as the Lorentzian) distribution which is a bell-shaped distribution whose "tails" are "longer" or "fatter" than the normal. Figure 10d illustrates the probability density function for the Cauchy distribution. The typical nature of long-tailed distributions like the Cauchy is that if the measurement process is generating data from such a distribution, it is more likely to generate some observations which are considerably removed from the "body" of the data than in sampling from a more moderate-tailed distribution (such as the normal). As before, since the original normal probability plot was not linear, the analyst should perform the iterative analysis to produce a longer-tailed probability plot (like a Cauchy probability plot). If this second plot is linear, this implies that the Cauchy yields an acceptable distribution. If this second plot is not linear, other iterations on D_0 must be made based on the S-shaped or N-shaped appearance of the Cauchy probability plot. A routine computerized procedure to carry out such iterations for the symmetric family of distributions will be presented in section 11.

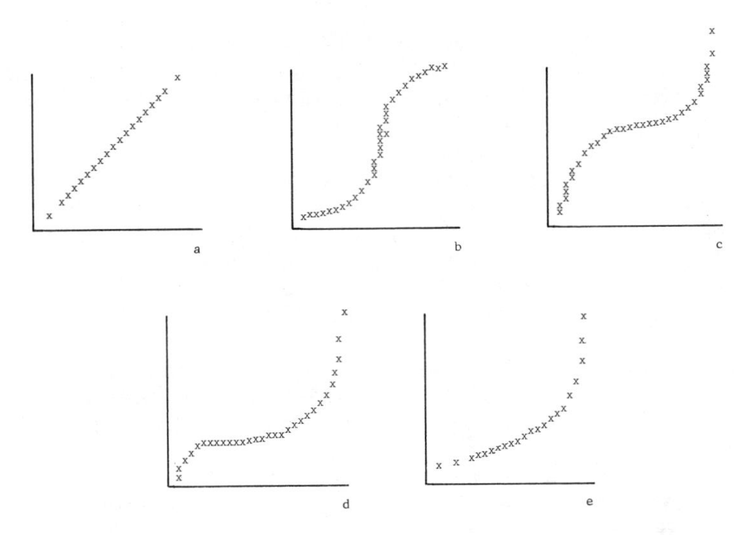

Figure 12. Typical shapes of probability plots. (a.) Linear; (b.) s-shaped; (d.) non-symmetric crossover; (e.) convex.

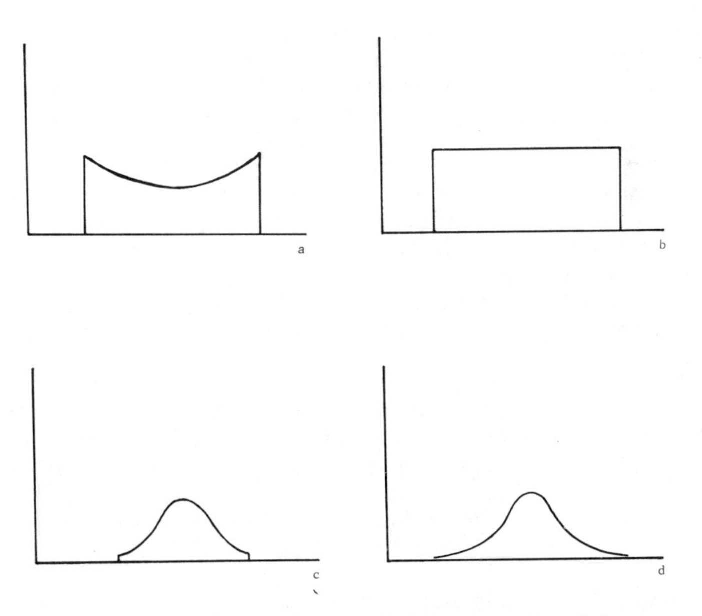

Figure 13. Distributions shorter-tailed than normal. a. Tukey λ = 1.5 distribution (very short-tailed); b. uniform distribution (short-tailed); c. truncated normal distribution (moderate/short-tailed); d. normal distribution (moderate-tailed).

If the original normal probability plot has the appearance of figure 12d where the diagonal line divided the data points on either side unequally or as in 12e where the diagonal line does not divide the data at all, this is indicative that not only may the specific hypothesis that D_0 = normal, be incorrect, but also that the hypothesis of a symmetric distribution may be incorrect. In such a case, the true underlying distribution for the data would then be some type of skewed distribution (e.g., of the types with probability density functions as illustrated in figure 11). In forming additional probability plots to fit the data, the analyst should consequently consider distributions which are skewed. To enumerate but a few of the skewed distributions that might be considered in subsequent iterations, one would include the log-normal distribution, the half-normal distribution, the exponential distribution, the Weibull family of distributions, the extreme value family of distributions, and the Pareto family of distributions. For an excellent general description of various distributions and distributional families (both skewed and symmetric) the reader is referred to the comprehensive texts by Johnson and Kotz (22,23).

One final point regarding outlier-detection is noteworthy. If in forming, for example, a normal probability plot, the plot turns out to be linear with the exception of one or two points (see fig. 14), how is this to be interpreted? This type of plot is indicating that the normal fit is acceptable for most of the data but that one or two points are outliers and do not seem to agree with the normality assumption. The probability plot is thus seen to be usable for detecting outliers. The next step in the analysis is for the analyst to delete the one or two offending points and to form a probability plot with the remaining points. If this second plot is still strongly linear, this gives additional support to the hypothesis that the data are normally-distributed and that the one or two questionable points are in fact outliers. The use of the probability plot as a tool for outlier detection is generally more sensitive than any of the techniques discussed in previous sections. The experimenter is also reminded that although such outliers may be deleted from further analysis, these outliers exist "for a reason" and the experimenter ought to satisfy himself that he has determined what set of experimental circumstances had led to them. The examination of outliers almost invariably leads to improved design of the experiment and ultimately to an improved understanding of the experimental factors which prevent a measurement process from being "in control."

Having discussed what a probability plot is and how one is to be interpreted, we now enumerate briefly some of the advantages of using a probability plot as opposed to other methods of checking for distributional information (e.g., histogram, χ^2 statistic, fit to probability density function).

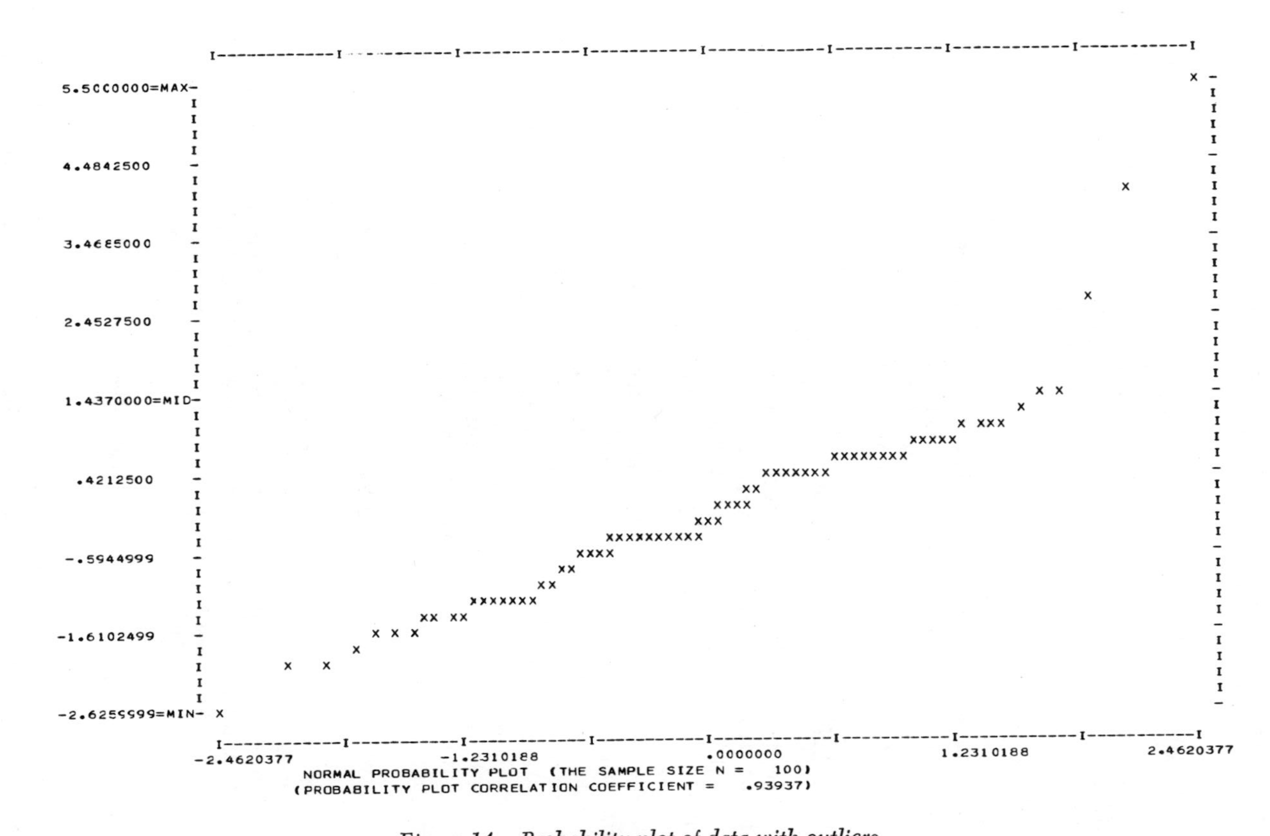

Figure 14. Probability plot of data with outliers

Although it will be shown that the probability plot technique is to be highly recommended, the various techniques are complementary. An outline of the advantages of the probability plot approach is as follows:

Graphical Technique

The probability plot is a graphical technique and so benefits from all of the advantages of graphics as outlined at the end of section 1.

Easy to Use

The dominant feature to be checked in a probability plot is linearity. This is the simplest structure possible in a plot and is easily detectable. In addition, the generation of probability plots is no longer a problem.

Applicable to a Wide Range of Distribution

The probability plot technique can be applied to a wide range of distributions--certainly for all distributions commonly encountered in practice. These distributions would cover those of both the continuous (*e.g.*, normal) and the discrete (*e.g.*, Poisson) types. Such distributions would include the normal (Gaussian), uniform, various U-shaped distributions, Cauchy, Logistic, half-normal, log-normal, exponential, gamma, beta, Weibull, extreme value, Pareto, binomial, Poisson, geometric, and negative binomial. For each such distribution D, there nonetheless remains the same uniform approach in interpreting the resulting probability plot; *viz.*, to check for linearity and if nonlinear to make adjustments to the hypothesized distributions D_0 accordingly--based on the type of nonlinearity encountered.

No *a priori* Location and Variation Estimates Needed

One problem associated with the χ^2 goodness of fit techniques and with the empirical technique of superimposing a fitted probability density function over a histogram of the data is that *a priori* values of the parameter (usually location and variation) are needed *before* the technique can actually be applied. This is frequently impractical for two reasons:

1. Such known values for the parameters are rarely available.

2. Accurate estimates for the parameters can only be obtained after the distribution has been "estimated" rather than *before*.

Since the probability plot technique does not need a priori values to be applied, it is superior and definitely far more practical than the χ^2 and fitted probability density function methods for distributional testing.

Automatic Estimate of Location and Variation Obtained

An additional advantage of applying the technique is that estimates of location and scale parameters are automatically produced as a secondary output. These location and variation estimates are derivable, respectively, from the vertical axis intercept and the slope of the resulting probability plot. Although the analyst is reminded that such location and variation estimates are not to be considered as the optimal (minimum variance) estimates, they nevertheless serve as useful practical indications of what the appropriate parameter values should be.

No Grouping of Data Need be Done

A problem associated with the histogram technique (whereby the analyst simply forms a histogram of the data and notes its general shape without applying or fitting a specific distribution to it) for gathering distributional information is that of choosing the grouping interval (the class width) for the histogram. The appearance of the resulting histogram is rather strongly affected by the choice of this class width. A class width which is "too narrow" will result in a histogram in which the true distributional shape is obscured by excessive variability in the height of the bar associated with each class. a class width which is "top wide" will result in a histogram in which the true distributional shape is obscured by "leakage" across neighboring classes so that the distributional content for a given class will be "smeared" out over several classes. Although rules of thumb do exist for choosing a reasonable class width, this nevertheless calls for an intermediate judgment to be made by the analyst. The use of the probability plot technique eliminates the need for such a choice. Inasmuch as a probability plot uses each observation individually and requires no grouping, this frees the analyst from making choices about class widths and eliminates (if the wrong class width happens to have been chosen) a possible undesirable approach-dependency on the ultimate conclusions. The net positive effect of the probability plot is that it allows a distributional analysis to be performed in a completely direct and automatic fashion with no intermediate decisions (such as class width) to be made by the analyst. Thus, the conclusions from the distributional analysis will reflect only the content of the data and will avoid possible biases introduced by the analysis.

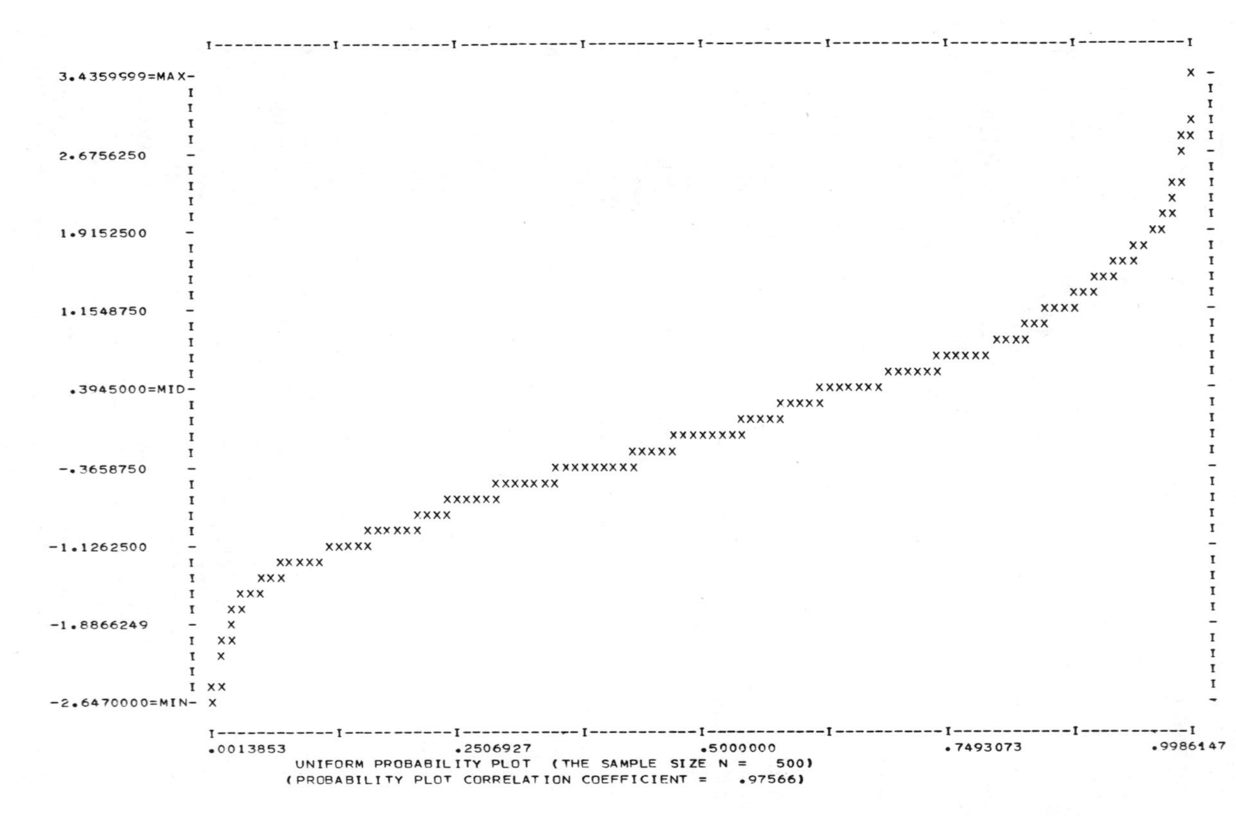

Figure 15a. Probability plots for Rand normal random deviates. Uniform (short-tailed) probability plot.

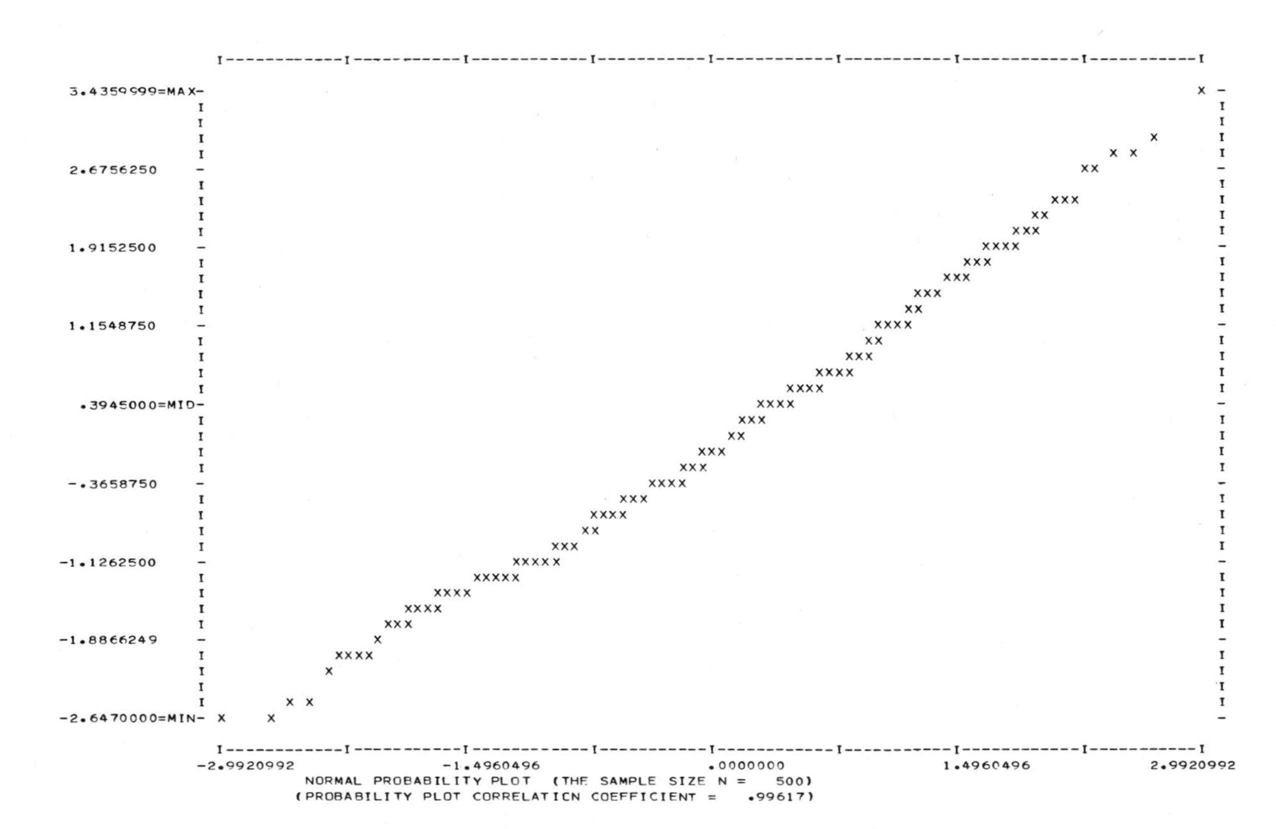

Figure 15b. Probability plots for Rand normal random deviates. Normal (moderate-tailed) probability plot.

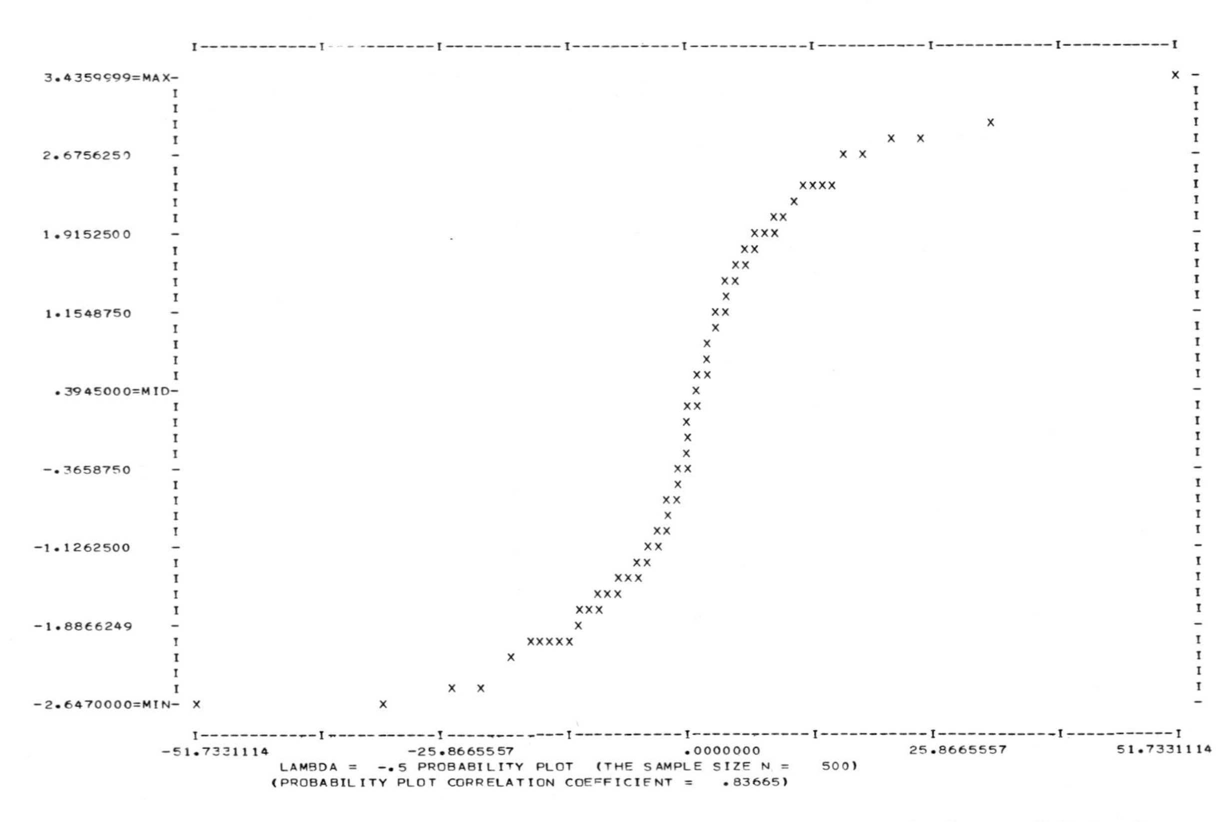

Figure 15c. Probability plots for Rand normal random deviates. Tukey λ = −.5 (moderate-tailed) distribution.

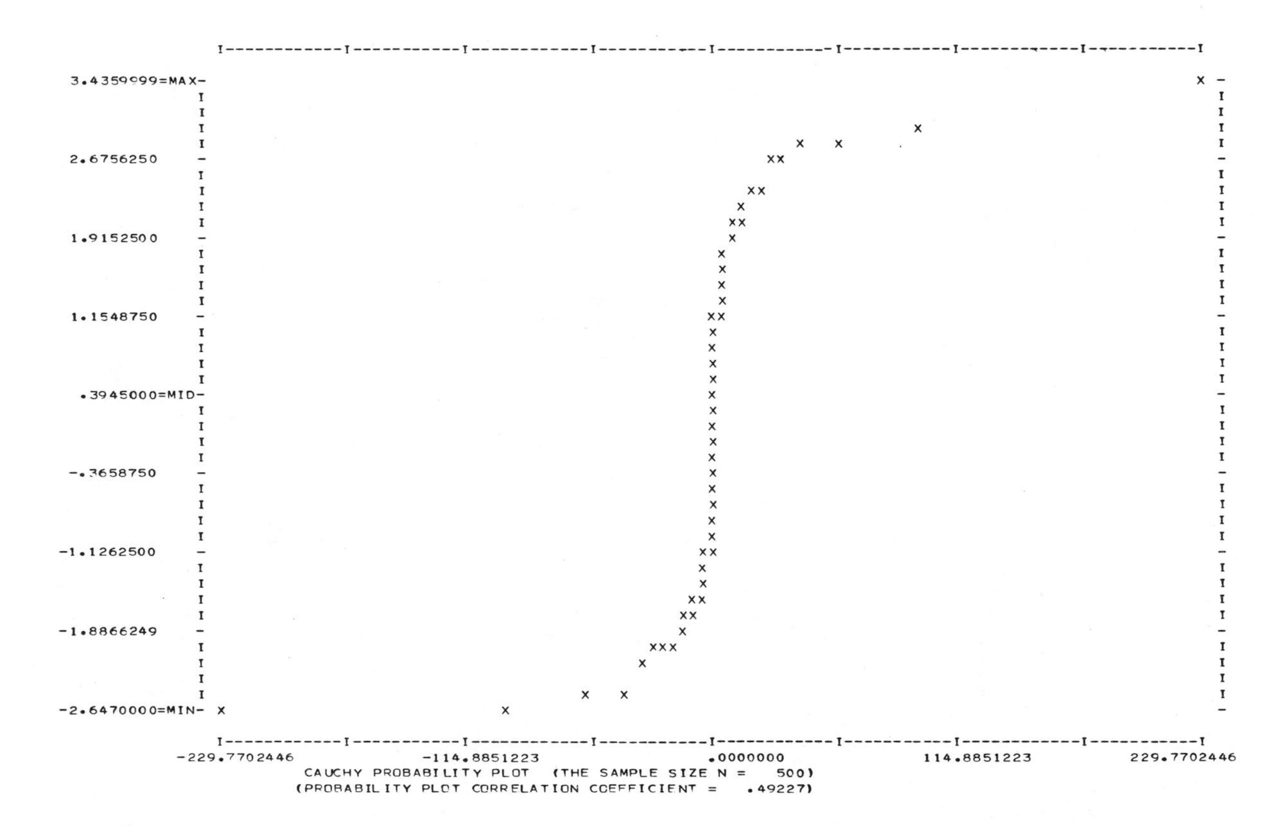

Figure 15d. Probability plots for Rand normal random deviates. Cauchy (long-tailed) probability plot.

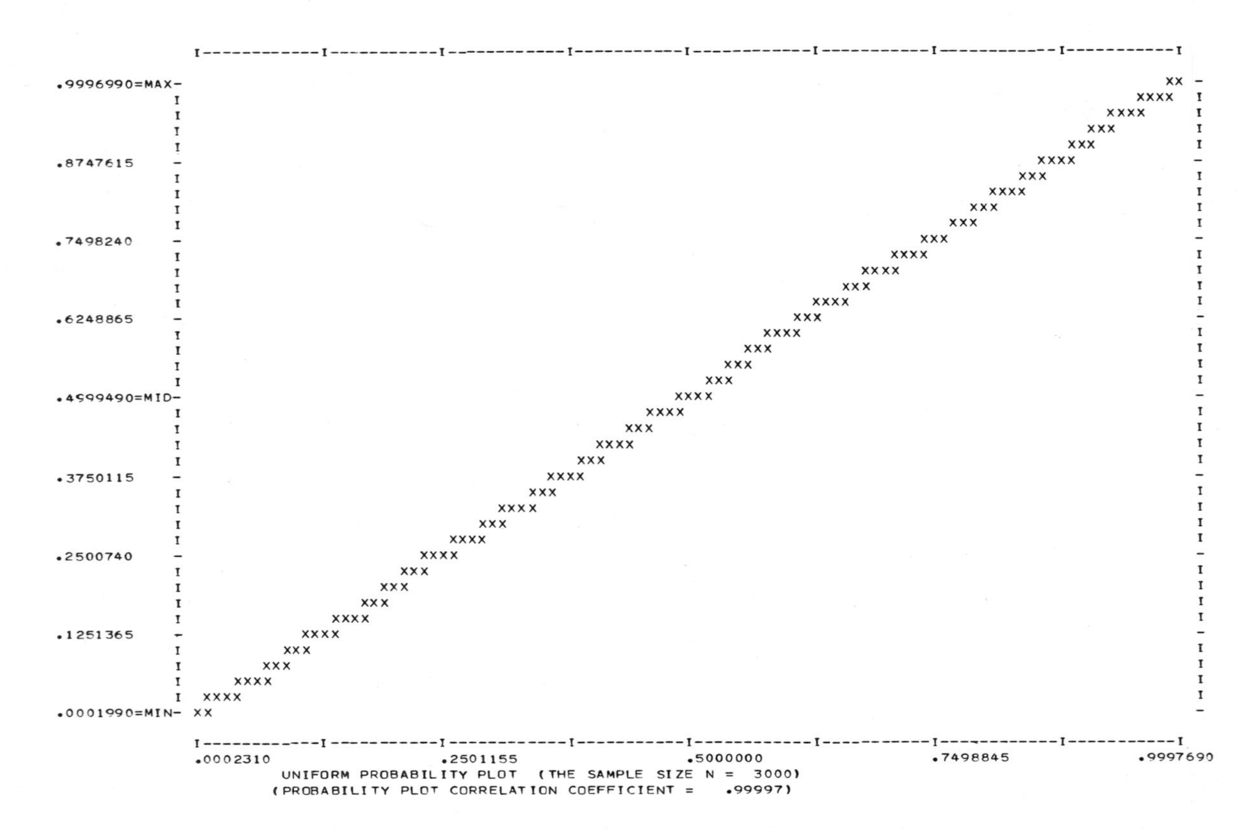

Figure 16a. Probability plots for Rand uniform random numbers. Uniform.

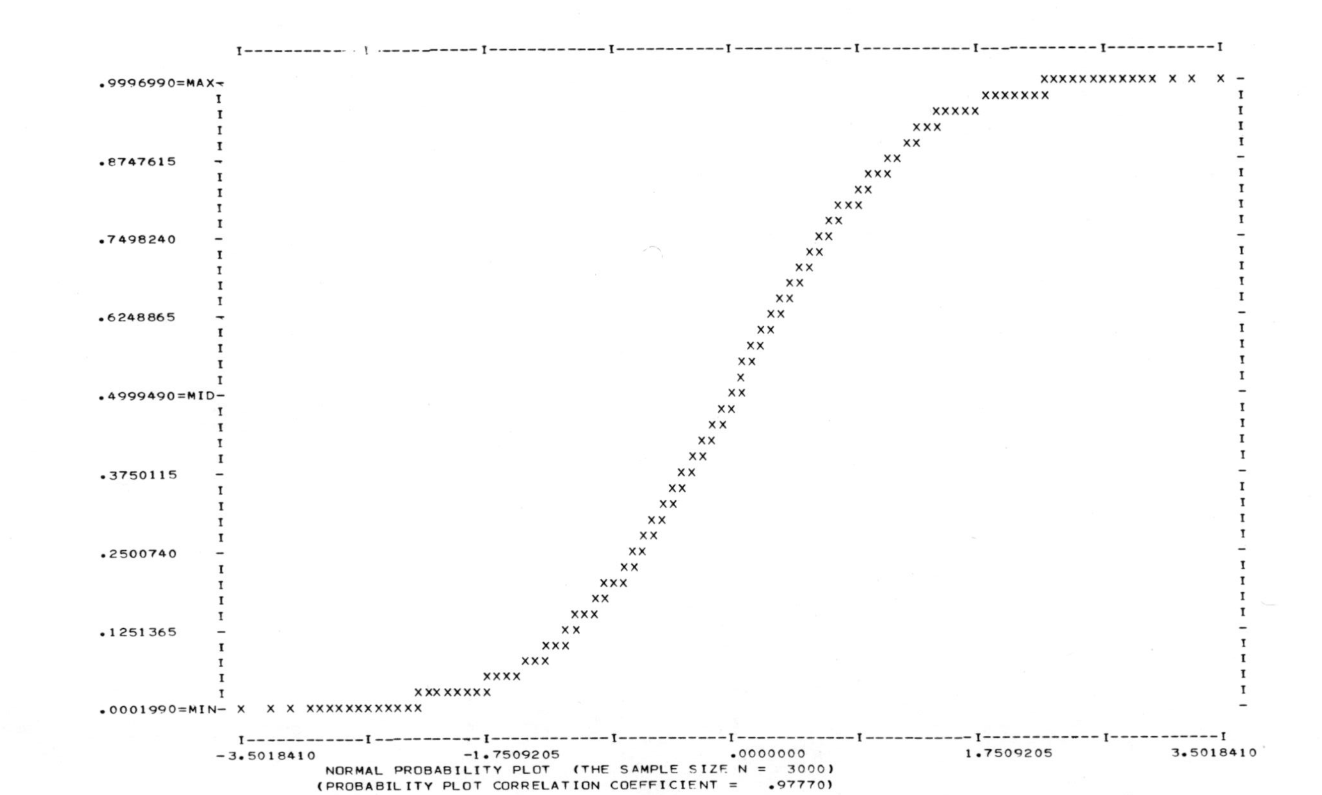

Figure 16b. Probability plots for Rand uniform random numbers. Normal.

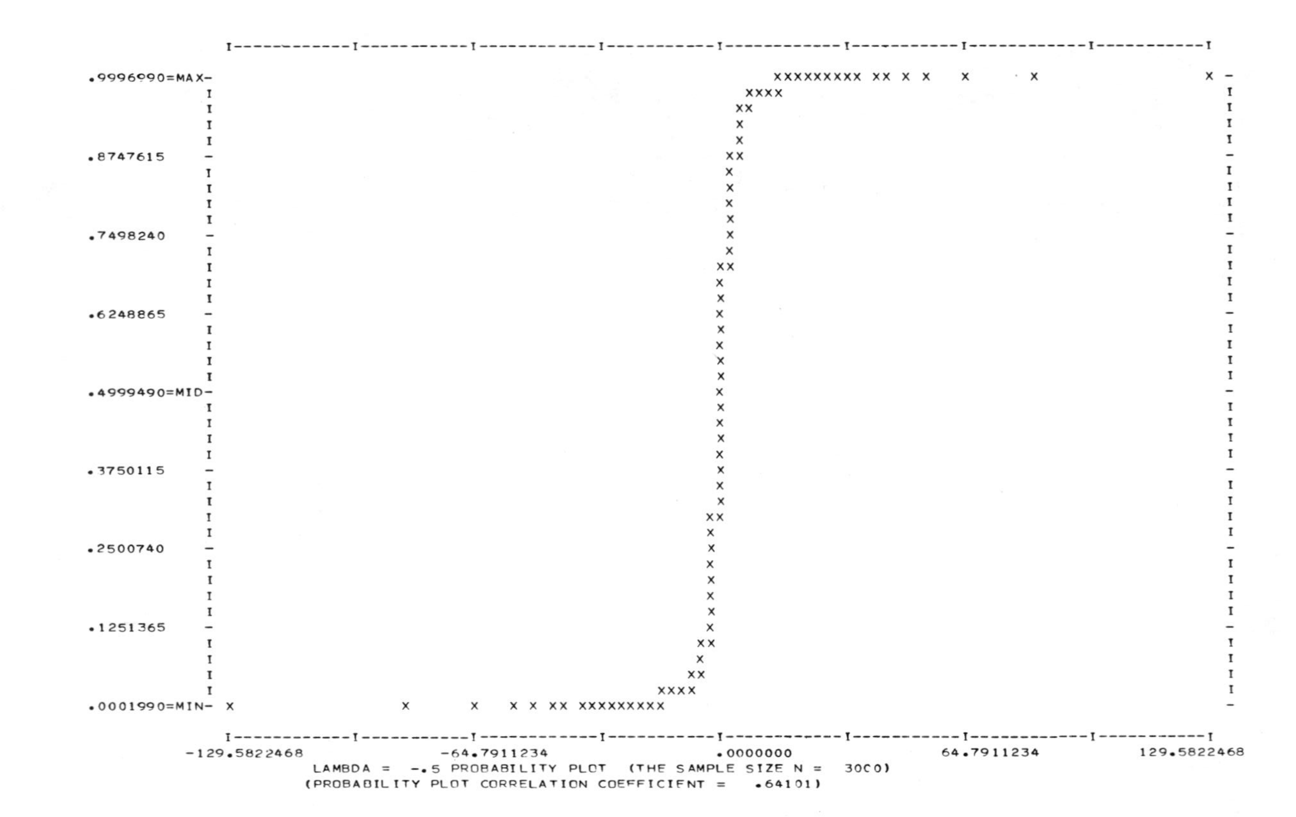

Figure 16c. Probability plots for Rand uniform random numbers. Tukey $\lambda = -.5$.

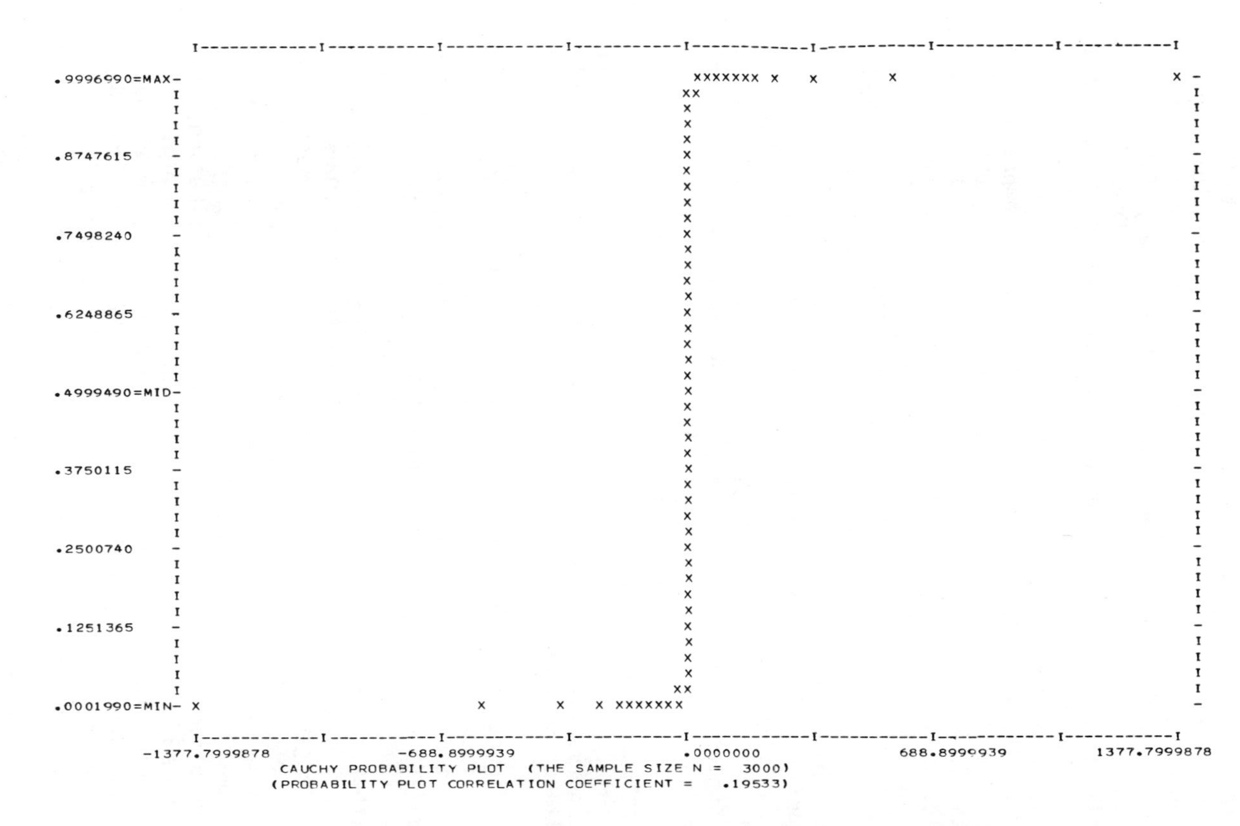

Figure 16d. Probability plots for Rand uniform random numbers. Cauchy.

Obtain Feedback Information for Improved Distributional Fit

Any statistic by its very nature attempts to map information (about some characteristic of interest) latent in the entire data set into a single number. Needless to say, such a mapping must be very selective (in order to be sensitive) and so invariably a considerable amount of ancillary information will be lost in the mapping. The above is true for test statistics in general and for any distributional test statistic in particular. Thus, for example, the standardized third central moment may be sensitive to symmetry but is not sensitive to tail length; the standardized 4th central moment may be sensitive to tail length but not to symmetry; and the χ^2 statistic may be sensitive to the distribution being tested, but yields no information on how to improve the distributional fit in the event of a poor distributional fit by the original tested distribution. Although not all distributional test statistics are equally bad, they are all necessarily "deficient" in the sense that information is irretrievably lost in the reduction of n numbers (the n observations) into one number (the test statistic value). This negative feature of test statistics is avoided in general by graphical procedures which utilize all of the individual data points. Thus, in the event that the original distribution, D_0 does not yield a good fit to the data, the analyst is able to improve the distributional fit. As discussed in detail previously, the shape of a given probability plot automatically gives the analyst the feedback information necessary to make another iteration in attempting to improve the distributional fit.

Figures 15 through 21 demonstrate the use of the probability plot on various data sets. The first example is a data set consisting of 500 normal random numbers drawn from the Rand (24) Corporation random number tables. Even though it is known that the true D_0 is normal, probability plots based on four hypothetical choices of distribution are given in figure 15. These four distributions (drawn from different regions of the tail length domain) are given for comparative purposes. The upper left plot is a uniform (short-tailed) probability plot, the upper right is a normal (moderate-tailed) probability plot, the lower left is a probability plot for the Tukey $\lambda = -.5$ distribution (25,26) (a moderate-long tailed distribution), and the lower right is a Cauchy (long-tailed) probability plot. Note the linear character of the normal probability plot as it should be since the considered data set is normal by construction. Note also the characteristic N- and S-shapes of the other three probability plots.

The second data example (fig. 16) consists of 500 random numbers generated from a uniform distribution. These random numbers were also based on the Rand (24) Corporation random number tables. Again, probability plots for the same four distributions are presented for comparative purposes. Note how the uniform probability plot is most linear (as it should be) and how the other three probability plots are decidedly nonlinear (as they should be). The above two data examples were included to illustrate the sensitivity of the probability plot technique.

The third example (fig. 17) is the 700 voltage readings collected from the Josephson Junction cryothermometry experiment described in sections 2 and 3. Note how (even through the inherent discreteness of the data) it is seen that the normal distribution provides the best fit of the four distributions considered.

The fourth example (fig. 18) is the 1200 wind velocity readings. The normal distribution also provides the best fit for this data set, although the hump in the normal line is suggestive of a location and correlation problem as has already been discussed for this data set in sections 2 and 3.

The fifth example (fig. 19) is the 200 steel-concrete beam deflection readings when subjected to a periodic pressure. Note how the four probability plots are all S-shaped, and that the best fit (of the four hypothesized distributions) is provided by the uniform distribution. Note also how the uniform fit is obviously not fully linear; so another iteration is appropriate, in the shorter-tailed direction since the plot is S- rather than N-shaped. As it turned out, a short-tailed, U-shaped distribution resulted in a probability plot that was quite linear and so provided a much better fit to the data than any of the four distributions considered in figure 19.

Example 6 (fig. 20) is residuals from a rather complicated 100-variable least squares fit of x-ray crystallography data. None of the four probability plots is linear. From the nature of the transition between the N-shaped probability plots (uniform and normal) the S-shaped probability plots (Tukey λ = -.5 distribution), there is seen a suggestion that the largest three data points are outliers. A subsequent examination of these three data points revealed that they were collected quite early in the experiment and so may possibly be reflecting an instrument warm-up effect.

Example 7 (fig. 21) is 53 yearly maximum wind speeds (27,28) for Valentine, Nebraska. The four usual symmetric distributions were all nonlinear in the fashion of 12d and 12e, and so suggested that some skewed distributions might better fit the data. The probability plot given is for one such skewed distribution--the log-normal distribution. It is seen that the probability plot is roughly linear but not optimally so and some other skewed distribution might provide an improved distribution fit.

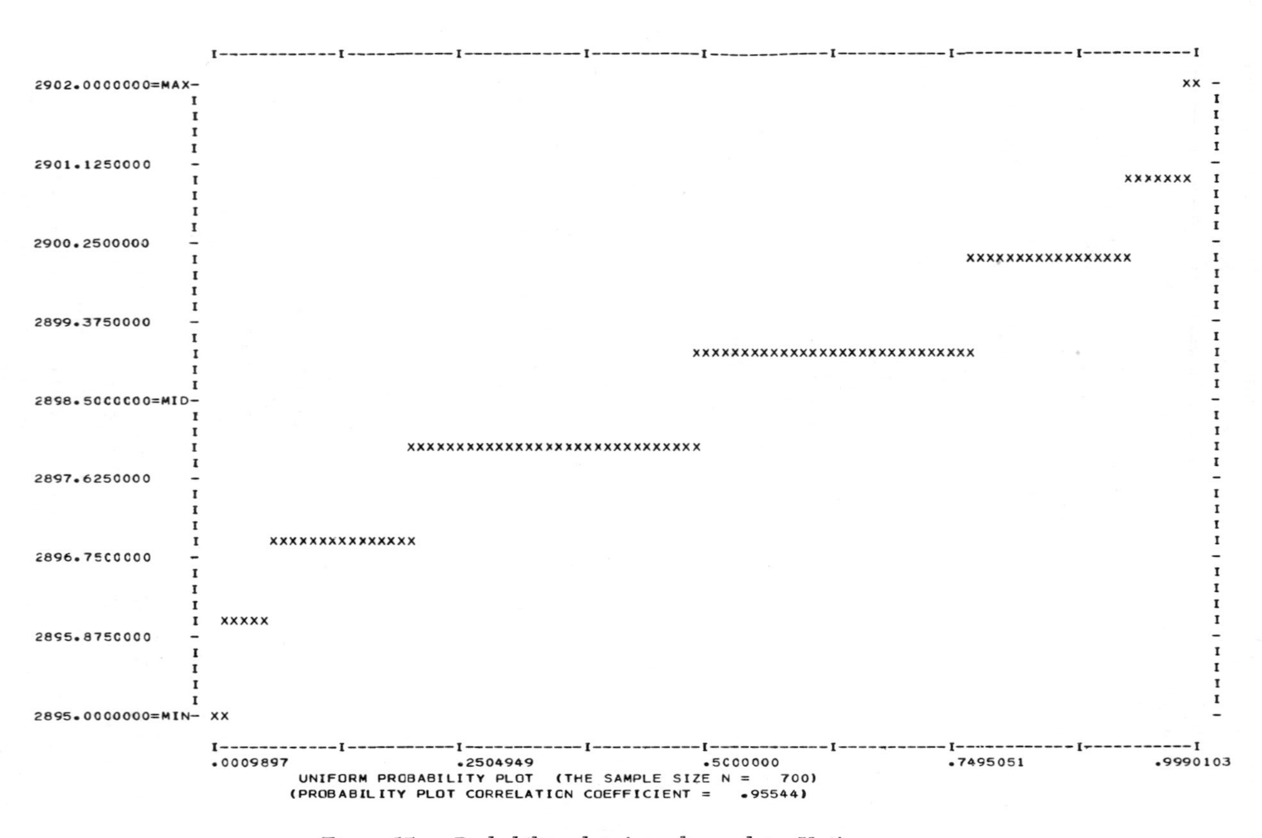

Figure 17a. Probability plots for voltage plots. Uniform.

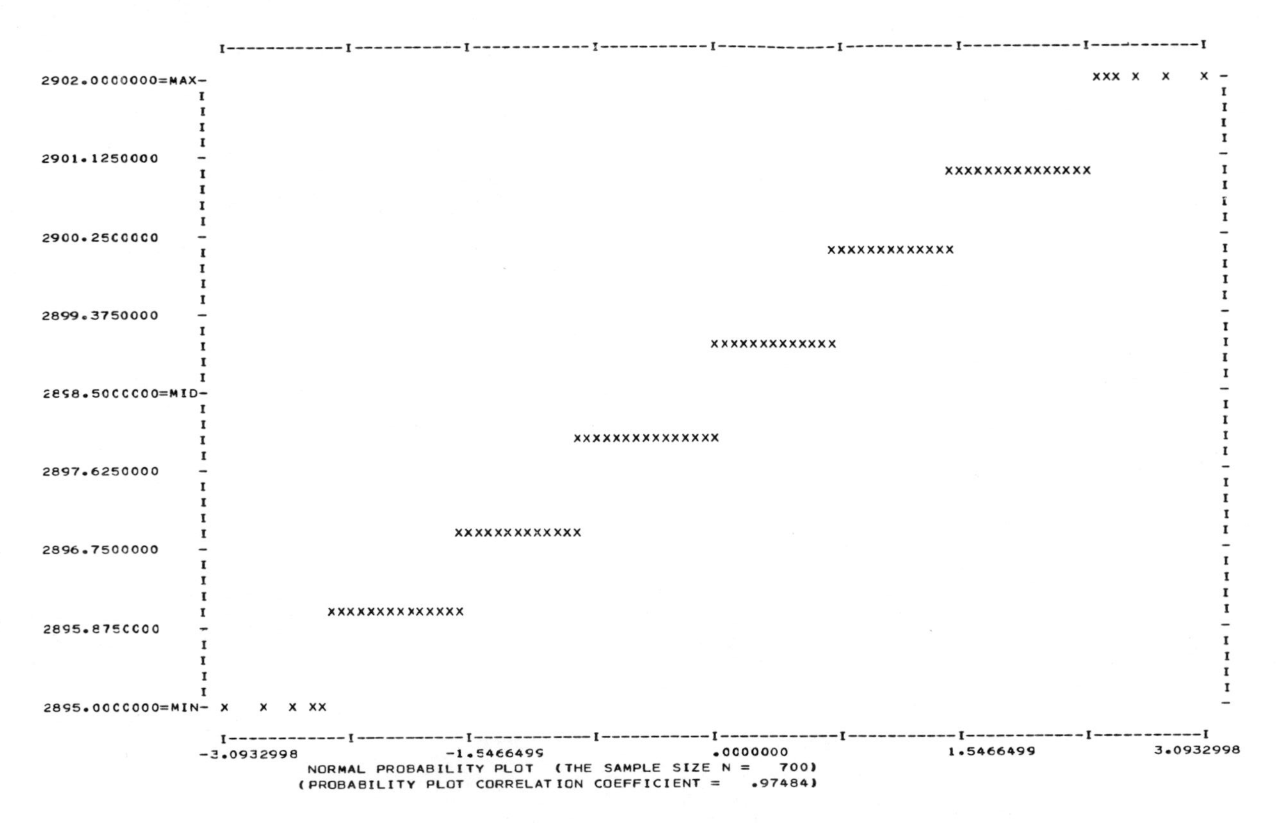

Figure 17b. Probability plots for voltage counts. Normal.

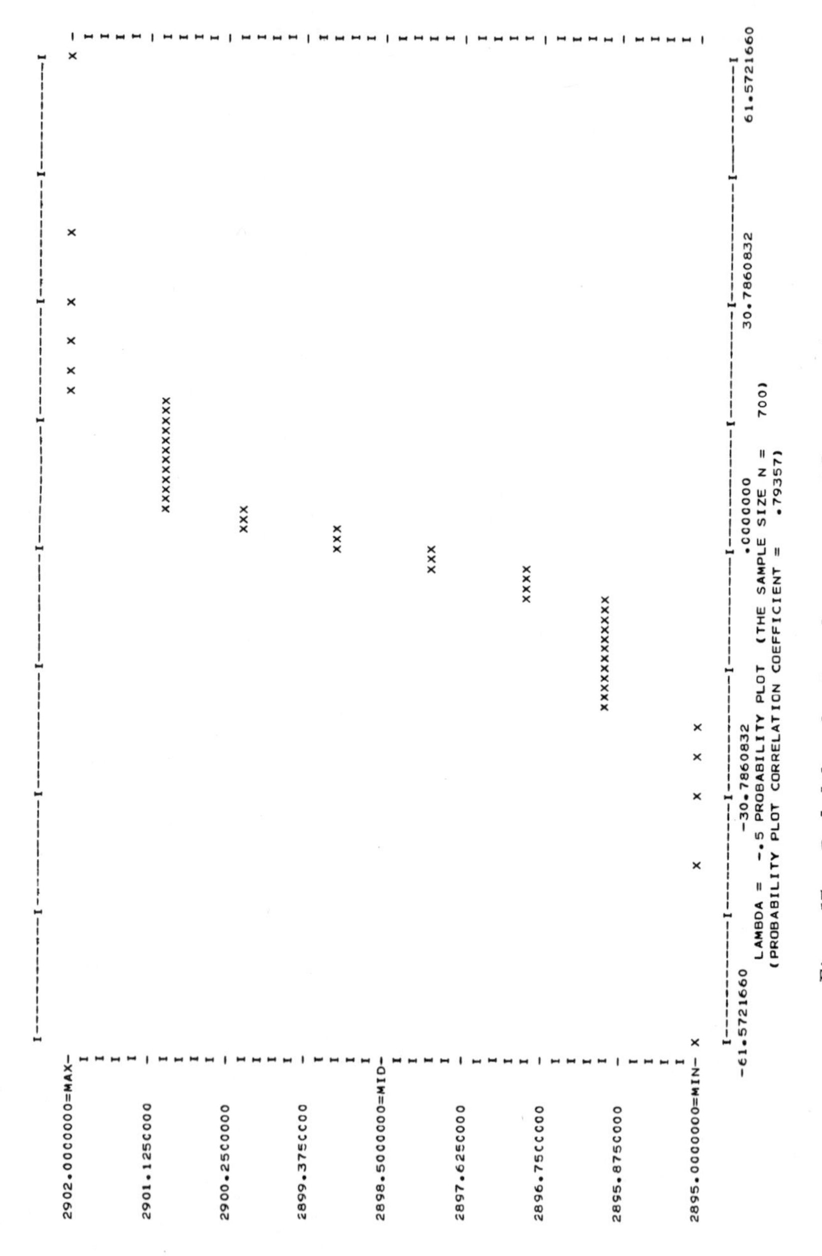

Figure 17c. Probability plots for voltage counts. Tukey $\lambda = -.5$.

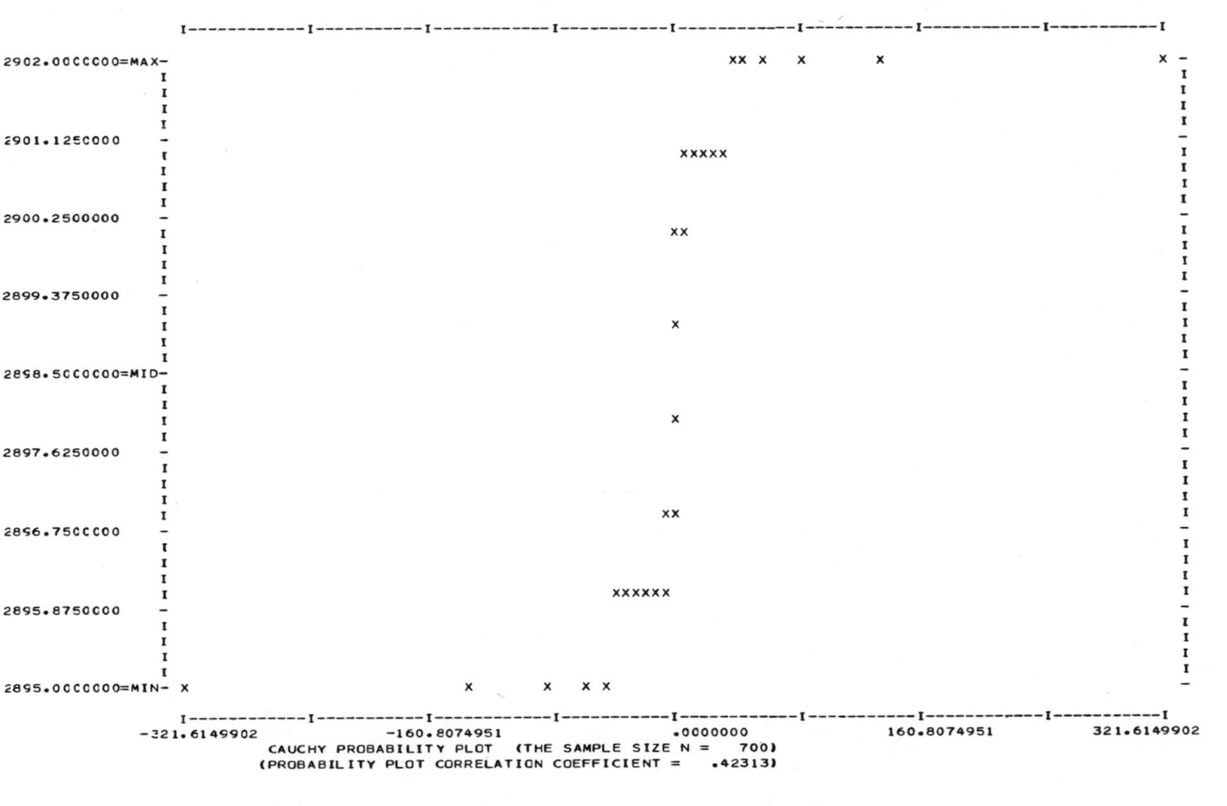

Figure 17d. Probability plots for voltage counts. Cauchy.

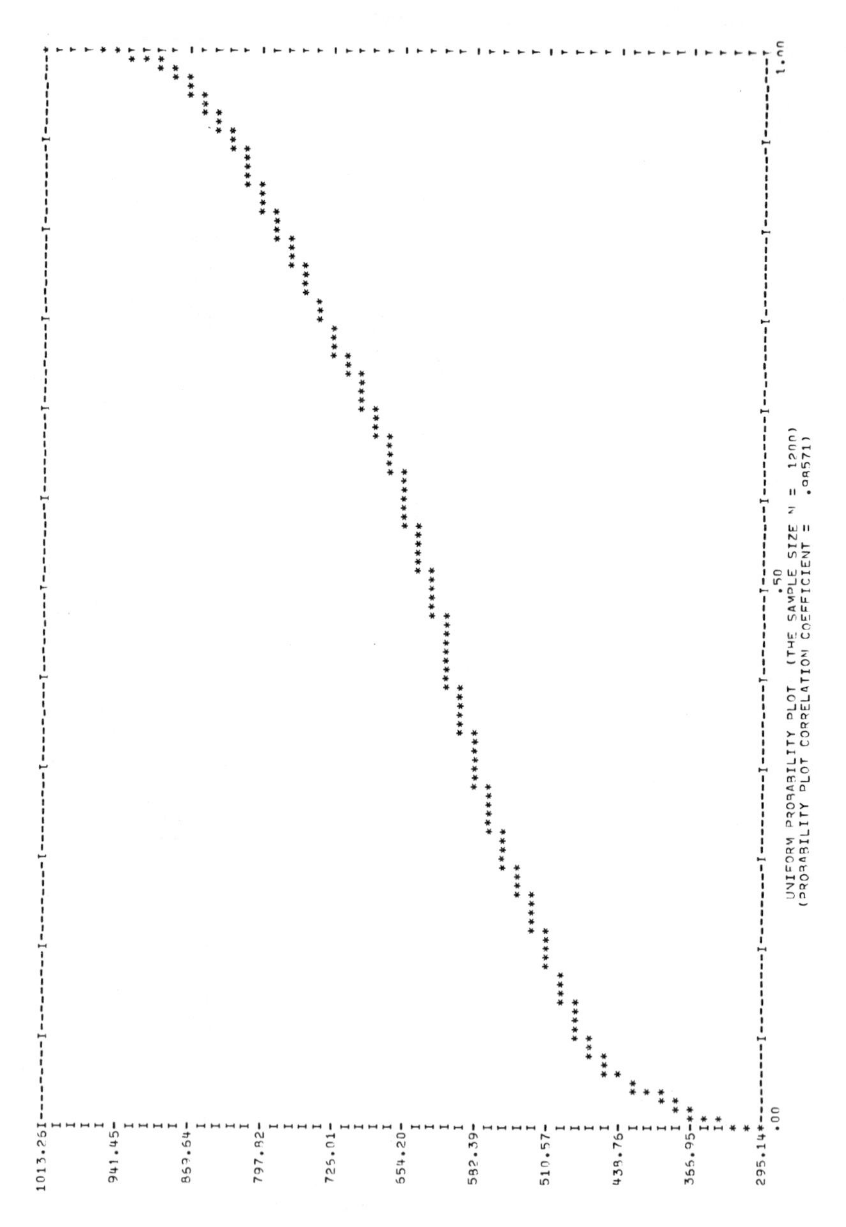

Figure 18a. Probability plots for wind velocities. Uniform.

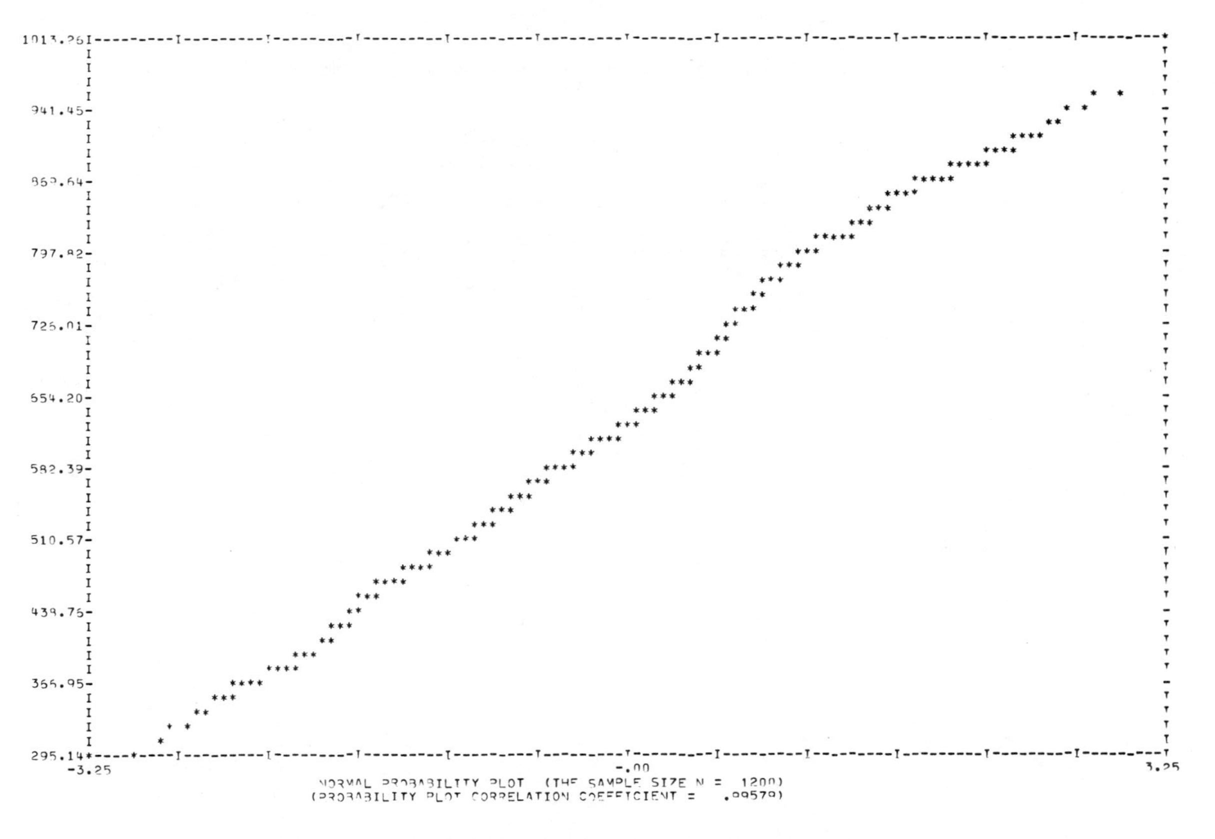

Figure 18b. Probability plots for wind velocities. Normal.

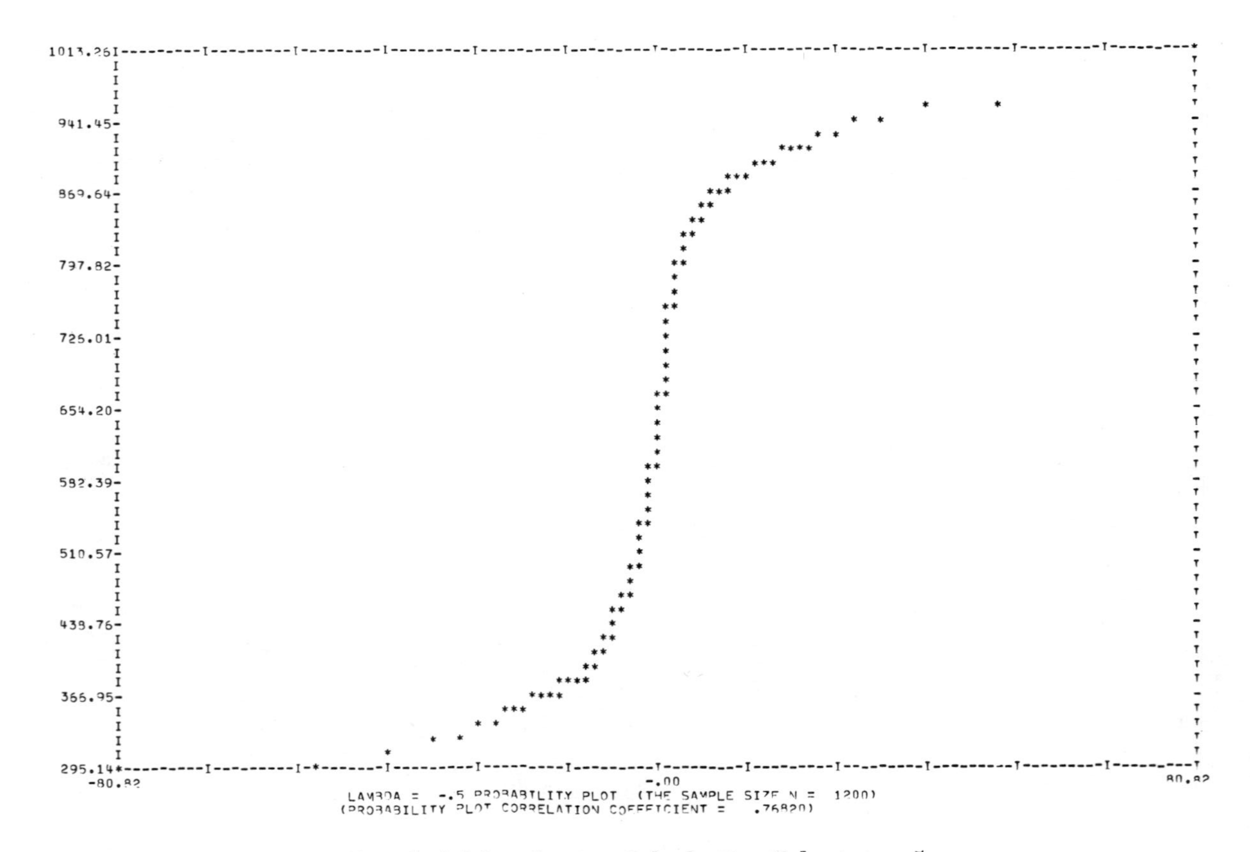

Figure 18c. Probability plots for wind velocities. Tukey $\lambda = -.5$.

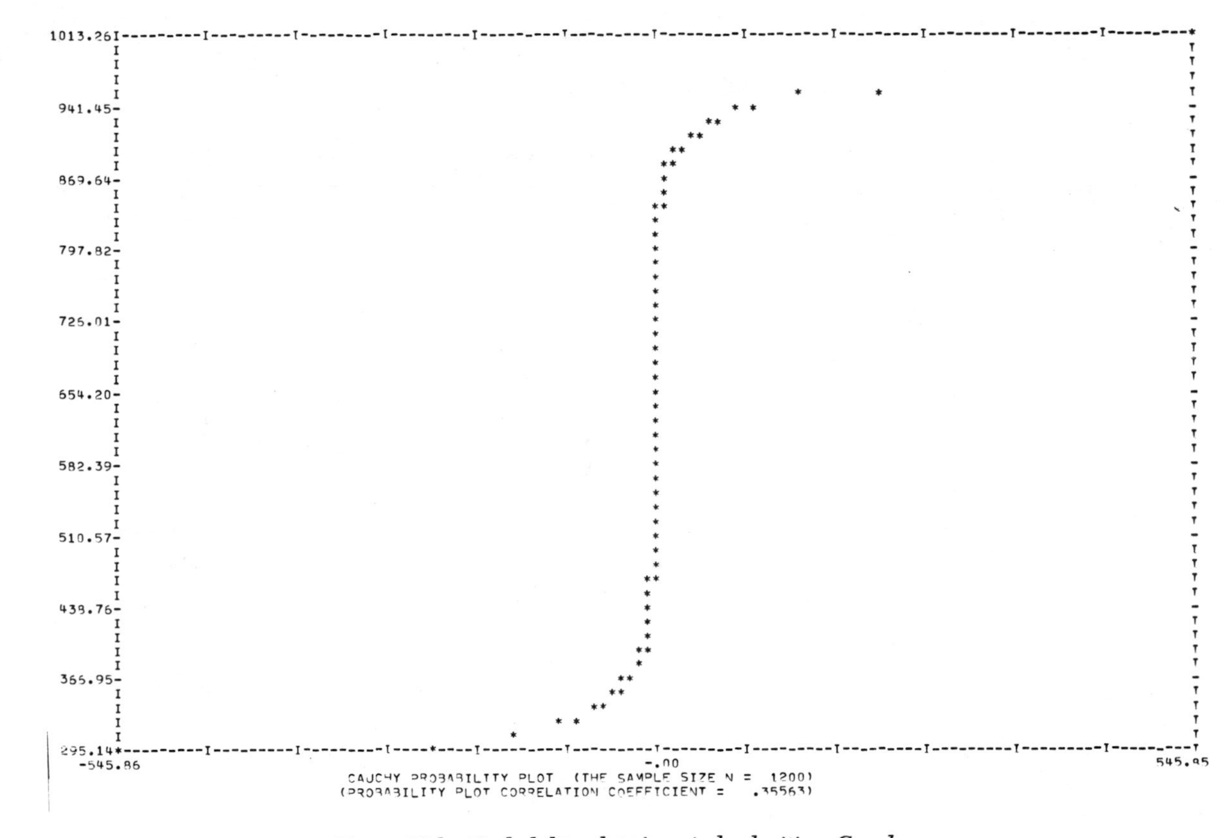

Figure 18d. Probability plots for wind velocities. Cauchy.

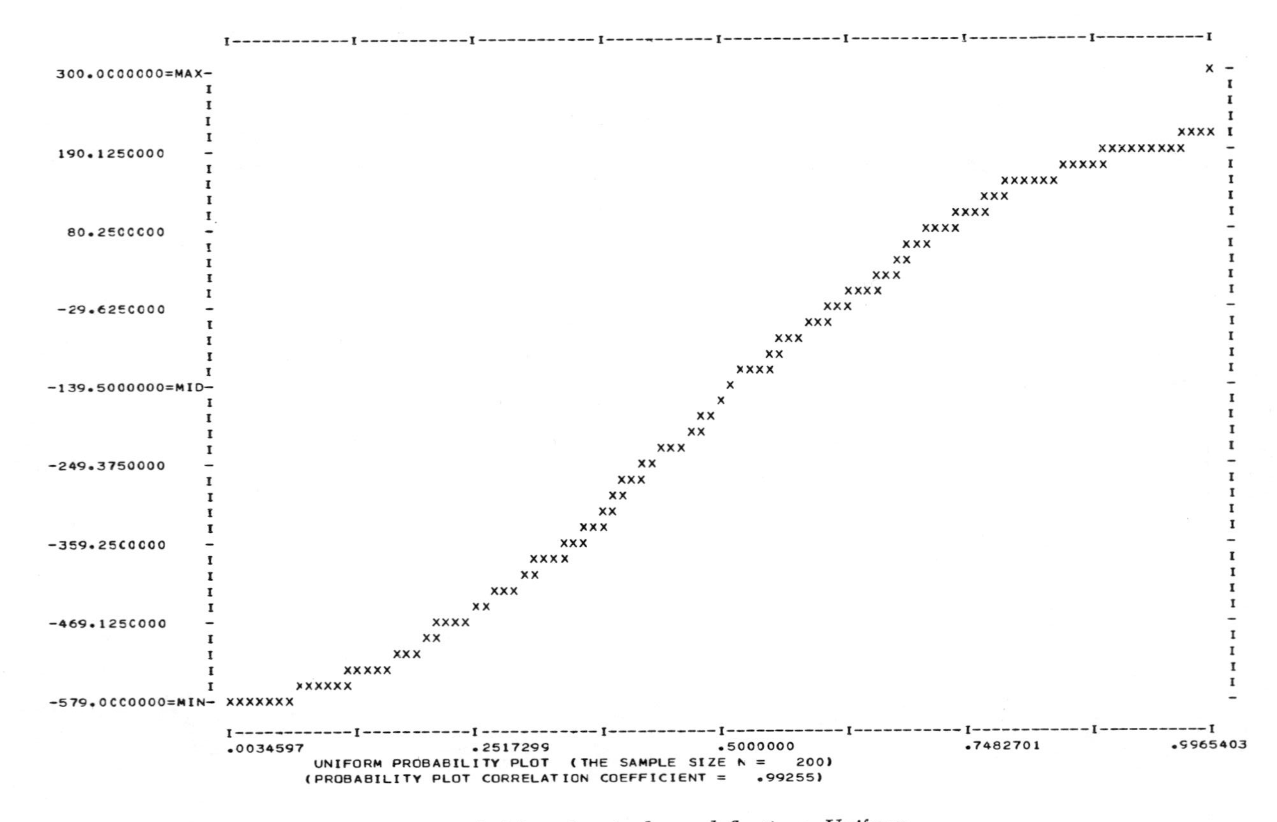

Figure 19a. Probability plots for beam deflections. Uniform.

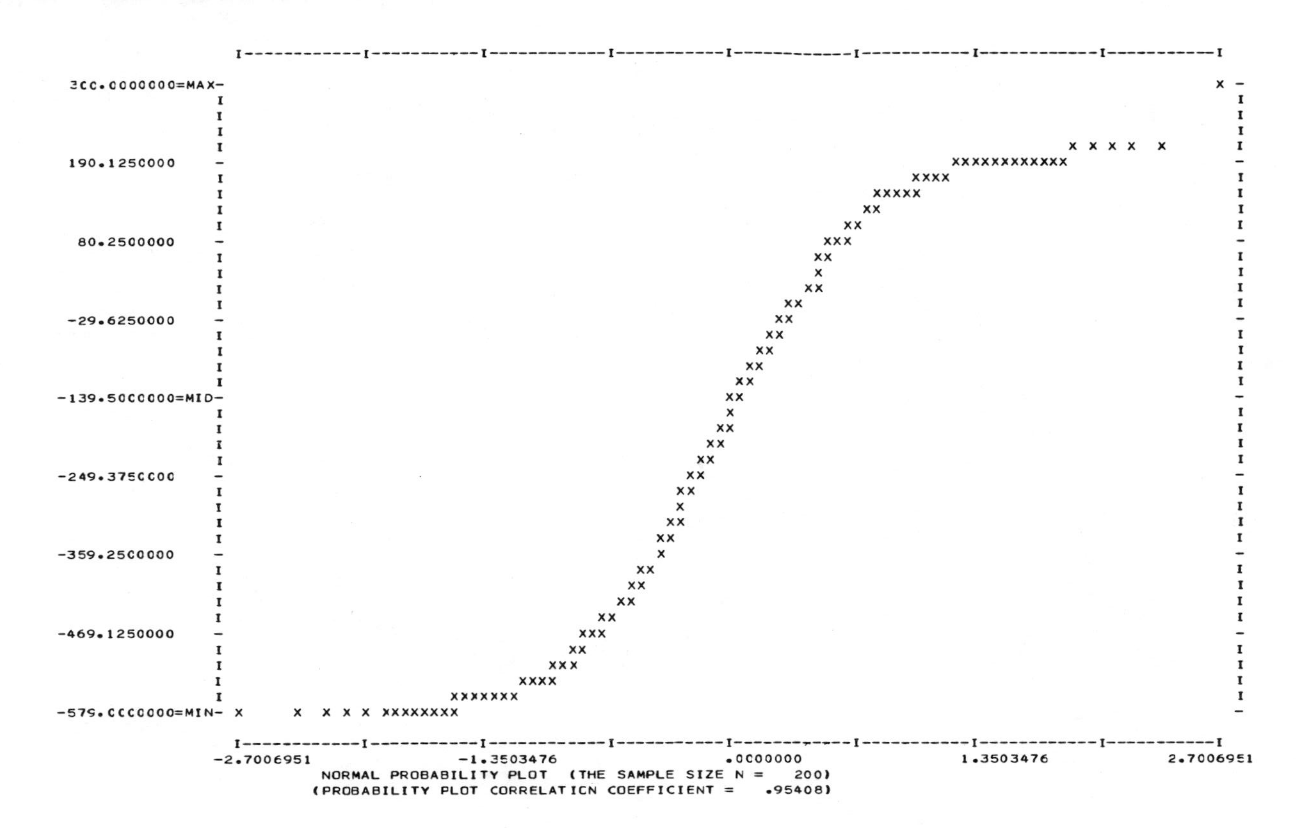

Figure 19b. Probability plots for beam deflections. Normal.

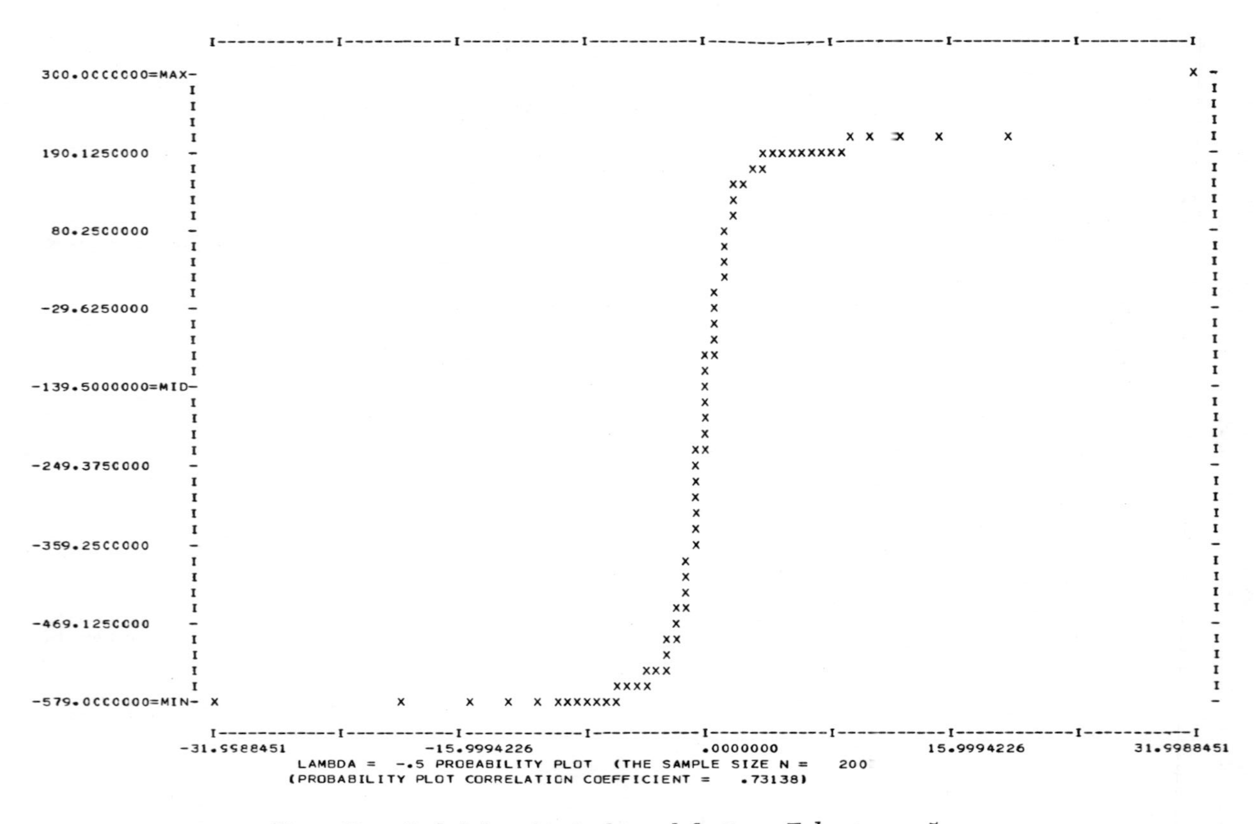

Figure 19c. Probability plots for beam deflections. Tukey $\lambda = -.5$.

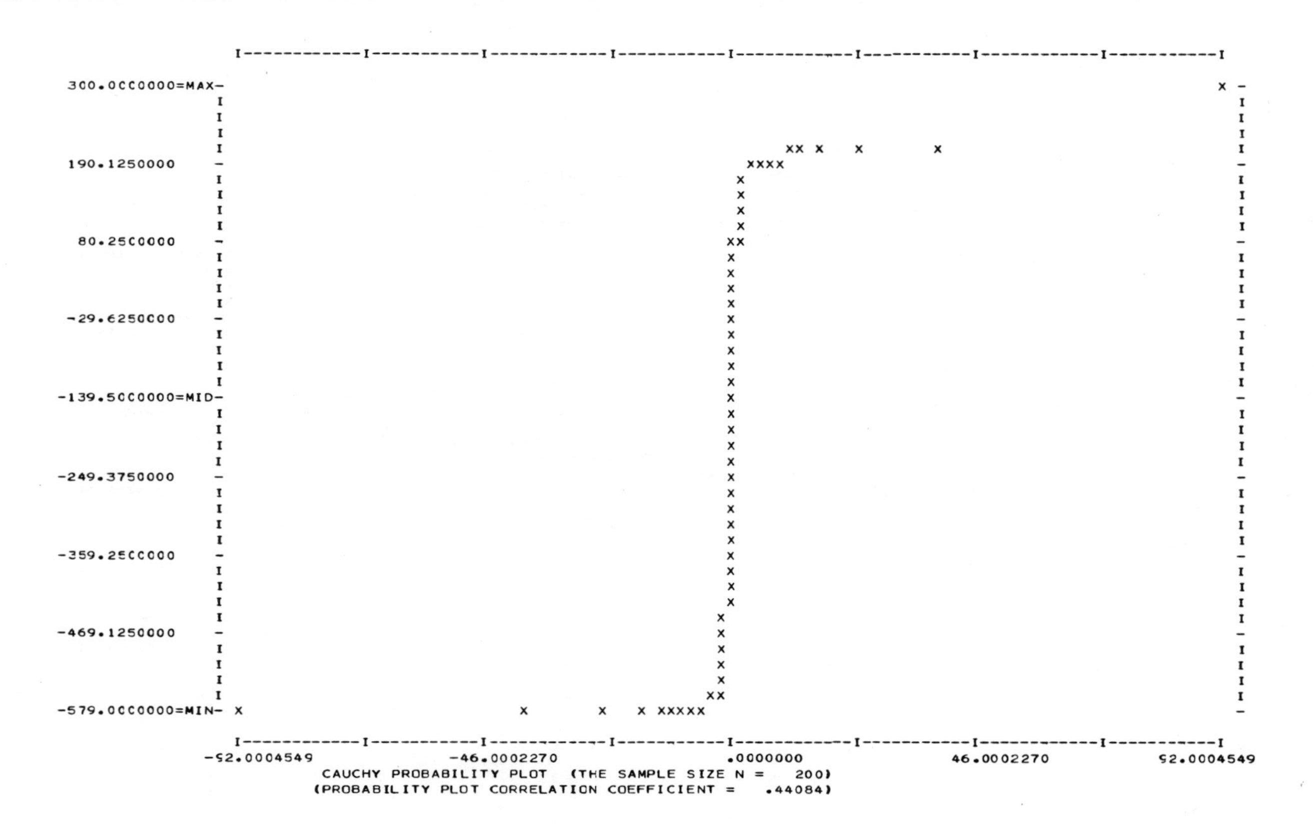

Figure 19d. Probability plots for beam deflections. Cauchy.

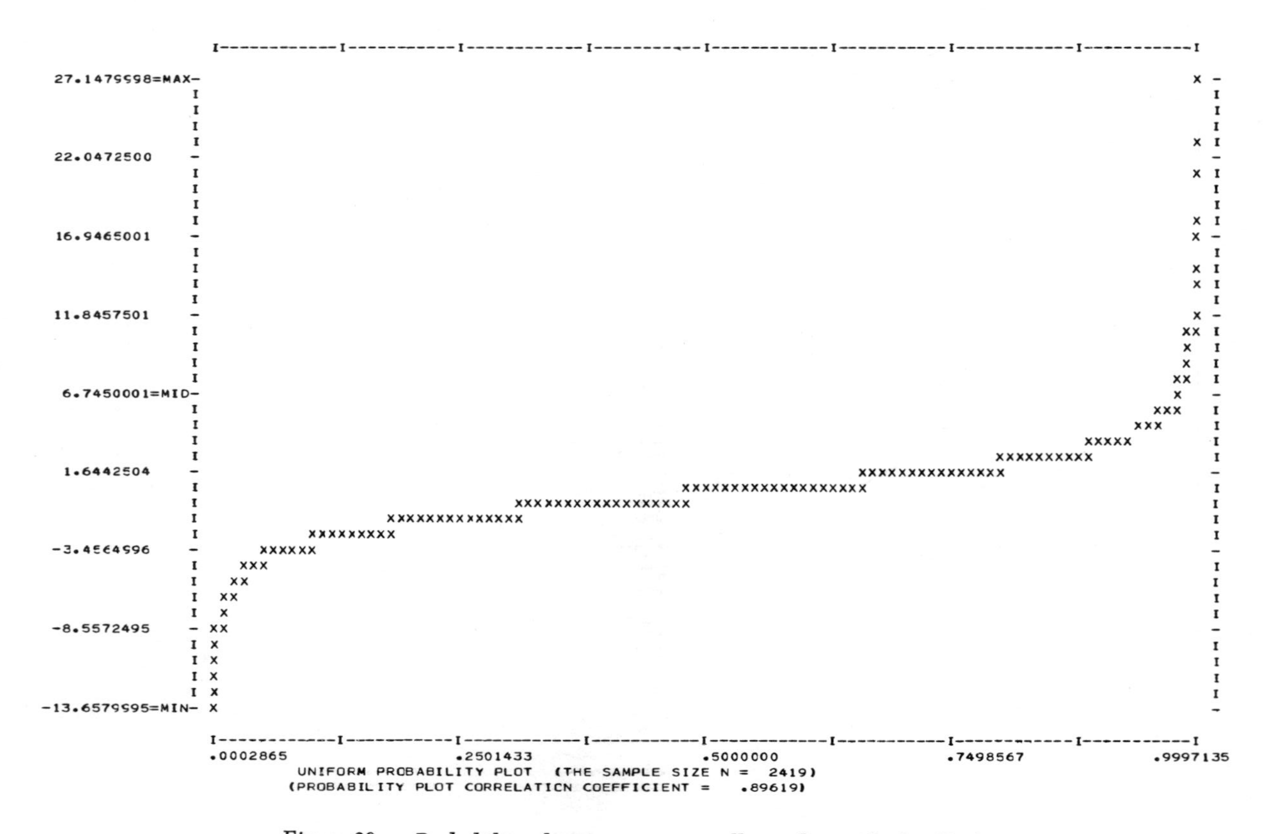

Figure 20a. Probability plots for x-ray crystallography residuals. Uniform.

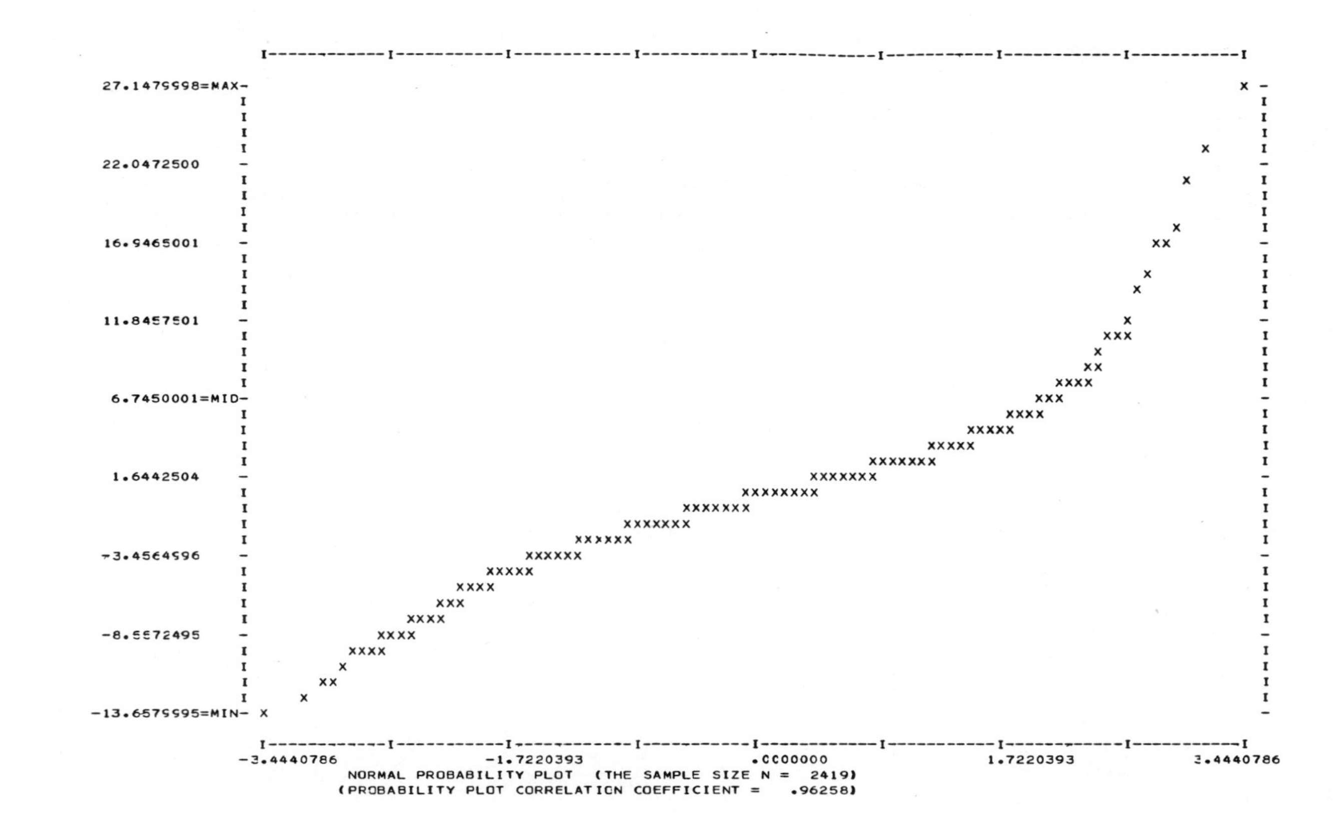

Figure 20b. Probability plots for x-ray crystallography residuals. Normal.

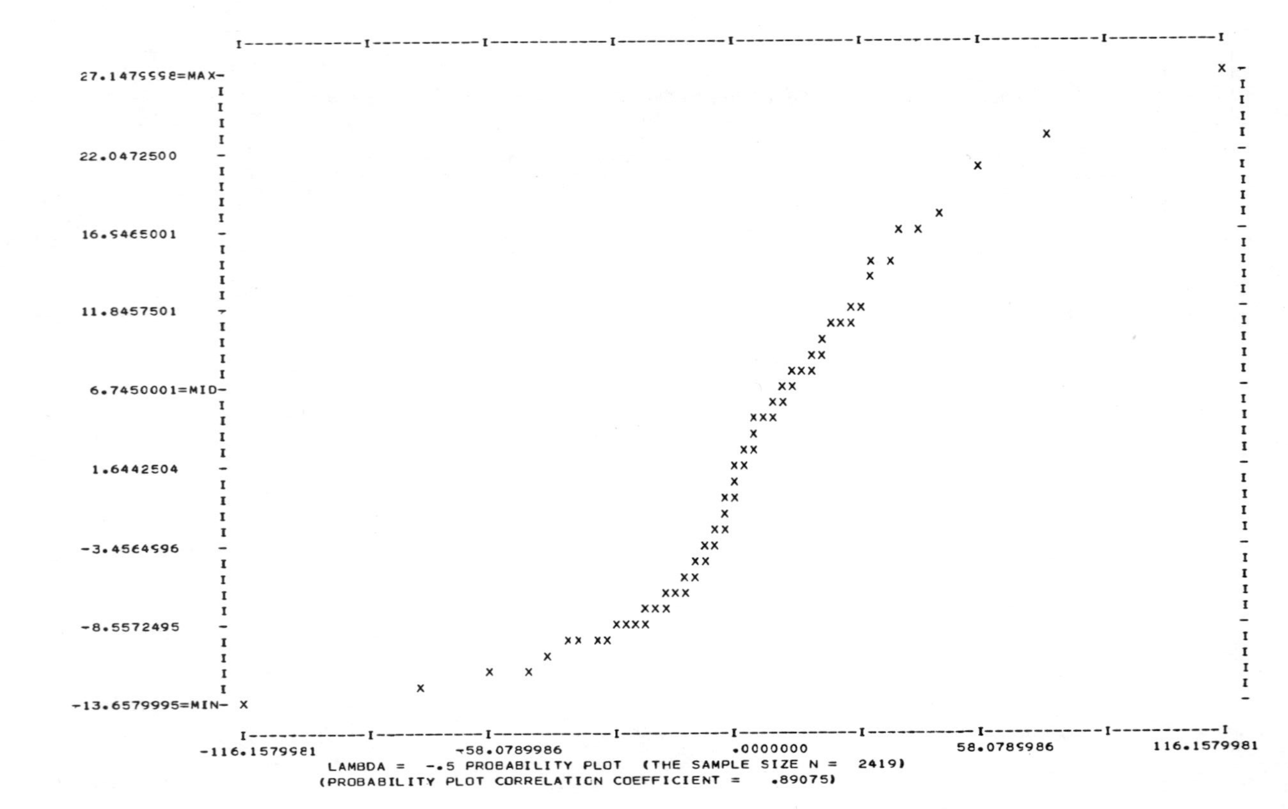

Figure 20c. Probability plots for x-ray crystallography residuals. Tukey $\lambda = -.5$.

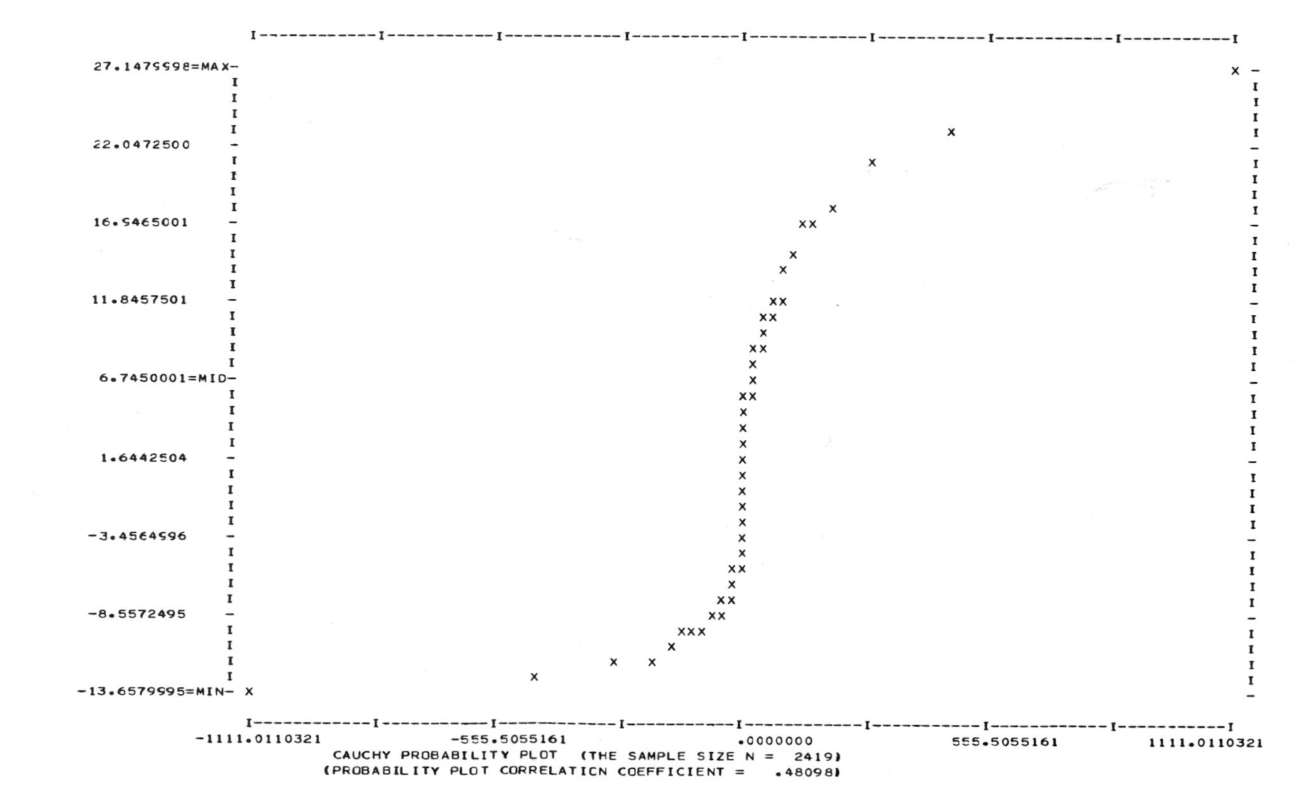

Figure 20d. Probability plots for x-ray crystallography residuals. Cauchy.

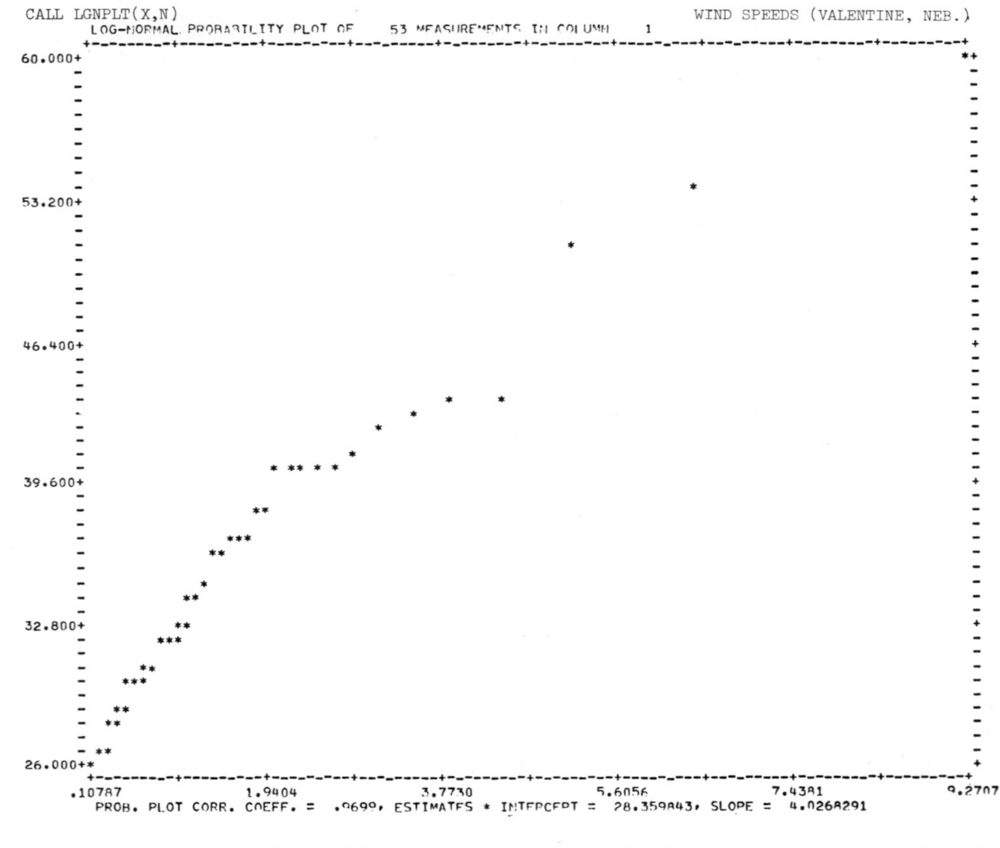

Figure 21. Log-normal probability plot for Valentine, Nebraska annual maximum wind speeds

PROBABILITY PLOT CORRELATION COEFFICIENT

A natural extension of the probability plot concept is that of the probability plot correlation coefficient (14,15). The probability plot correlation coefficient is an attempt to summarize the linearity information latent in a probability plot into a single statistic. A natural choice for such a linearity statistic is the product moment correlation coefficient, r which in general terms (for any two variables X and Y) is defined as:

$$r = \mathrm{Corr}(X,Y) = \frac{\Sigma(X_i - X)(Y_i - Y)}{\sqrt{\Sigma(X_i - X)^2 \ \Sigma(Y_i - Y)^2}}$$

where

$$X = \frac{\Sigma Y_i}{n} \quad \text{and} \quad Y = \frac{\Sigma Y_i}{n}$$

If the variables X and Y are linearly related in a positive fashion (i.e., if $Y = aX + b$ where $a > 0$), then $r = 1$. If the variables X and Y are linearly related in a negative fashion (i.e., if $Y = aX + b$ where $a < 0$), then $r = -1$ If the variables X and Y are unrelated, then $r \cong 0$. The stronger the linear dependency between X and Y, the closer r will be to +1 (or -1).

In a probability plot, the 2 "variables" of interest are the ordered observations Y and the "expected" (for a given distribution D_0) ordered observations M. The probability plot correlation coefficient is thus defined as $r = \mathrm{Corr}(Y, M)$. If the probability plot is near-linear, then r will be near-unity. If the probability plot is nonlinear, then r will be appropriately smaller than unity. The value of the probability plot correlation coefficient r, is thus a simple summary measure of the linearity of a given probability plot and hence a measure of the appropriateness of the distributional fit for the hypothesized distribution, D_0.

In practice the value of r for a single isolated distributional fit is not important in itself in an absolute sense. Of more importance is the relative values of r for various members of an admissible test set of distributions. If the four distributions (uniform, normal, Tukey $\lambda = -.5$, and Cauchy) from example 1 (RAND normal numbers--fig. 15) of the previous section were considered, and if the probability plot correlation coefficient for each of the four distributions would be much closer to unity than would the probability plot correlation coefficient for any of the other three distributions. Used in this fashion, one is led to consider the Maximum Probability Plot Correlation Coefficient (MPPCC) criterion as a reasonable one for choosing the "best" distribution out of a set of admissible distributions. The "best" as defined above will, of course, mean that distribution which yields the "most linear" probability plot.

Several examples are now presented which illustrate the use of the Maximum Probability Plot Correlation Coefficient criterion. Reverting to example 1 (Rand (24) normal random numbers) as described previously, the reader is directed to figure 22 which presents calculated values of the probability plot correlation coefficient for 44 selected symmetric distributions. These 44 distributions have been selected so as to present a dense tail-length coverage in the important symmetric distribution case. The λ values in the table are references to specific members of the Tukey, λ distribution family (25,26). The 44 distributions are selected to cover a wide range of tail lengths and are ordered from the very short-tailed distributions at top to the very long-tailed distributions at bottom. The values of λ between 2.0 and 1.0 represent very short-tailed U-shaped distributions, $\lambda = 1.0$ is identical to the uniform distribution. Values of λ smaller than 1 and larger than 0 are truncated bell-shaped distributions which are short to moderate-tailed in nature. $\lambda = 0$ is exactly the logistic distribution. Values of λ smaller than 0 represent infinite-domain bell-shaped distributions. The normal, logistic, double exponential and Cauchy distributions have been appropriately positioned in the tail-length ordering. The probability plot correlation coefficient values for each of the 44 hypothesized distributions is given on the far right of the table. Figure 22 is a single page which summarized the linearity information of 44 different symmetric distributional probability plots. Note how the probability plot correlation coefficients are smaller at the top and bottom and reach a maximum at the normal distribution (as it should since the data set was normal random numbers). The usefulness of such an analysis (which of course is completely computerized) is that the analyst can quickly determine not only the best-fitting distribution D_0 (in this case, normal), but also a subset of good-fitting distributions (in this case, distributions in the normal neighborhood) for the data set at hand. After narrowing the selection to an appropriate small neighborhood of distributions which have large values of the probability plot correlation coefficient, it is recommended that the analyst generate and scrutinize the actual probability plots for a few of the distributions in this neighborhood for outlier information and other possible anomalies.

As a second example, consider a data set consisting of 500 uniform random numbers based on the Rand (24) tables. The MPPCC criteria (see fig. 23) as applied to the uniform random numbers leads to the conclusion that the uniform distribution (as it should) provides the best distributional fit for these data.

```
THE CORRELATION BETWEEN THE  500 ORDERED OBS. AND THE ORDER STAT. MEDIANS FROM THE LAMBDA =  2.0 DIST. IS  .97566
THE CORRELATION BETWEEN THE  500 ORDERED OBS. AND THE ORDER STAT. MEDIANS FROM THE LAMBDA =  1.9 DIST. IS  .97482
THE CORRELATION BETWEEN THE  500 ORDERED OBS. AND THE ORDER STAT. MEDIANS FROM THE LAMBDA =  1.8 DIST. IS  .97407
THE CORRELATION BETWEEN THE  500 ORDERED OBS. AND THE ORDER STAT. MEDIANS FROM THE LAMBDA =  1.7 DIST. IS  .97347
THE CORRELATION BETWEEN THE  500 ORDERED OBS. AND THE ORDER STAT. MEDIANS FROM THE LAMBDA =  1.6 DIST. IS  .97302
THE CORRELATION BETWEEN THE  500 ORDERED OBS. AND THE ORDER STAT. MEDIANS FROM THE LAMBDA =  1.5 DIST. IS  .97277
THE CORRELATION BETWEEN THE  500 ORDERED OBS. AND THE ORDER STAT. MEDIANS FROM THE LAMBDA =  1.4 DIST. IS  .97275
THE CORRELATION BETWEEN THE  500 ORDERED OBS. AND THE ORDER STAT. MEDIANS FROM THE LAMBDA =  1.3 DIST. IS  .97299
THE CORRELATION BETWEEN THE  500 ORDERED OBS. AND THE ORDER STAT. MEDIANS FROM THE LAMBDA =  1.2 DIST. IS  .97353
THE CORRELATION BETWEEN THE  500 ORDERED OBS. AND THE ORDER STAT. MEDIANS FROM THE LAMBDA =  1.1 DIST. IS  .97441
THE CORRELATION BETWEEN THE  500 ORDERED OBS. AND THE ORDER STAT. MEDIANS FROM THE LAMBDA =  1.0 DIST. IS  .97566
THE CORRELATION BETWEEN THE  500 ORDERED OBS. AND THE ORDER STAT. MEDIANS FROM THE LAMBDA =   .9 DIST. IS  .97731
THE CORRELATION BETWEEN THE  500 ORDERED OBS. AND THE ORDER STAT. MEDIANS FROM THE LAMBDA =   .8 DIST. IS  .97937
THE CORRELATION BETWEEN THE  500 ORDERED OBS. AND THE ORDER STAT. MEDIANS FROM THE LAMBDA =   .7 DIST. IS  .98183
THE CORRELATION BETWEEN THE  500 ORDERED OBS. AND THE ORDER STAT. MEDIANS FROM THE LAMBDA =   .6 DIST. IS  .98465
THE CORRELATION BETWEEN THE  500 ORDERED OBS. AND THE ORDER STAT. MEDIANS FROM THE LAMBDA =   .5 DIST. IS  .98774
THE CORRELATION BETWEEN THE  500 ORDERED OBS. AND THE ORDER STAT. MEDIANS FROM THE LAMBDA =   .4 DIST. IS  .99089
THE CORRELATION BETWEEN THE  500 ORDERED OBS. AND THE ORDER STAT. MEDIANS FROM THE LAMBDA =   .3 DIST. IS  .99375
THE CORRELATION BETWEEN THE  500 ORDERED OBS. AND THE ORDER STAT. MEDIANS FROM THE LAMBDA =   .2 DIST. IS  .99575
THE CORRELATION BETWEEN THE  500 ORDERED OBS. AND THE ORDER STAT. MEDIANS FROM THE NORMAL DISTRIBUTION IS  .99617 MAX
THE CORRELATION BETWEEN THE  500 ORDERED OBS. AND THE ORDER STAT. MEDIANS FROM THE LAMBDA =   .1 DIST. IS  .99591
THE CORRELATION BETWEEN THE  500 ORDERED OBS. AND THE ORDER STAT. MEDIANS FROM THE LOGISTIC DIST.       IS  .99275
THE CORRELATION BETWEEN THE  500 ORDERED OBS. AND THE ORDER STAT. MEDIANS FROM THE DOUBLE EXP. DIST.    IS  .97897
THE CORRELATION BETWEEN THE  500 ORDERED OBS. AND THE ORDER STAT. MEDIANS FROM THE LAMBDA =  -.1 DIST. IS  .98409
THE CORRELATION BETWEEN THE  500 ORDERED OBS. AND THE ORDER STAT. MEDIANS FROM THE LAMBDA =  -.2 DIST. IS  .96699
THE CORRELATION BETWEEN THE  500 ORDERED OBS. AND THE ORDER STAT. MEDIANS FROM THE LAMBDA =  -.3 DIST. IS  .93815
THE CORRELATION BETWEEN THE  500 ORDERED OBS. AND THE ORDER STAT. MEDIANS FROM THE LAMBDA =  -.4 DIST. IS  .89487
THE CORRELATION BETWEEN THE  500 ORDERED OBS. AND THE ORDER STAT. MEDIANS FROM THE LAMBDA =  -.5 DIST. IS  .83665
THE CORRELATION BETWEEN THE  500 ORDERED OBS. AND THE ORDER STAT. MEDIANS FROM THE LAMBDA =  -.6 DIST. IS  .76637
THE CORRELATION BETWEEN THE  500 ORDERED OBS. AND THE ORDER STAT. MEDIANS FROM THE LAMBDA =  -.7 DIST. IS  .68997
THE CORRELATION BETWEEN THE  500 ORDERED OBS. AND THE ORDER STAT. MEDIANS FROM THE LAMBDA =  -.8 DIST. IS  .61424
THE CORRELATION BETWEEN THE  500 ORDERED OBS. AND THE ORDER STAT. MEDIANS FROM THE LAMBDA =  -.9 DIST. IS  .54457
THE CORRELATION BETWEEN THE  500 ORDERED OBS. AND THE ORDER STAT. MEDIANS FROM THE CAUCHY DISTRIBUTION IS  .49227
THE CORRELATION BETWEEN THE  500 ORDERED OBS. AND THE ORDER STAT. MEDIANS FROM THE LAMBDA = -1.0 DIST. IS  .48422
THE CORRELATION BETWEEN THE  500 ORDERED OBS. AND THE ORDER STAT. MEDIANS FROM THE LAMBDA = -1.1 DIST. IS  .43296
THE CORRELATION BETWEEN THE  500 ORDERED OBS. AND THE ORDER STAT. MEDIANS FROM THE LAMBDA = -1.2 DIST. IS  .39116
THE CORRELATION BETWEEN THE  500 ORDERED OBS. AND THE ORDER STAT. MEDIANS FROM THE LAMBDA = -1.3 DIST. IS  .35725
THE CORRELATION BETWEEN THE  500 ORDERED OBS. AND THE ORDER STAT. MEDIANS FROM THE LAMBDA = -1.4 DIST. IS  .32987
THE CORRELATION BETWEEN THE  500 ORDERED OBS. AND THE ORDER STAT. MEDIANS FROM THE LAMBDA = -1.5 DIST. IS  .30772
THE CORRELATION BETWEEN THE  500 ORDERED OBS. AND THE ORDER STAT. MEDIANS FROM THE LAMBDA = -1.6 DIST. IS  .28974
THE CORRELATION BETWEEN THE  500 ORDERED OBS. AND THE ORDER STAT. MEDIANS FROM THE LAMBDA = -1.7 DIST. IS  .27505
THE CORRELATION BETWEEN THE  500 ORDERED OBS. AND THE ORDER STAT. MEDIANS FROM THE LAMBDA = -1.8 DIST. IS  .26295
THE CORRELATION BETWEEN THE  500 ORDERED OBS. AND THE ORDER STAT. MEDIANS FROM THE LAMBDA = -1.9 DIST. IS  .25292
THE CORRELATION BETWEEN THE  500 ORDERED OBS. AND THE ORDER STAT. MEDIANS FROM THE LAMBDA = -2.0 DIST. IS  .24453
```

Figure 22. Printout plot correlation coefficient analysis for Rand normal random deviates

```
THE CORRELATION BETWEEN THE  3000 ORDERED OBS. AND THE ORDER STAT. MEDIANS FROM THE LAMBDA =  2.0 DIST. IS   .99997 MAX
THE CORRELATION BETWEEN THE  3000 ORDERED OBS. AND THE ORDER STAT. MEDIANS FROM THE LAMBDA =  1.9 DIST. IS   .99996
THE CORRELATION BETWEEN THE  3000 ORDERED OBS. AND THE ORDER STAT. MEDIANS FROM THE LAMBDA =  1.8 DIST. IS   .99992
THE CORRELATION BETWEEN THE  3000 ORDERED OBS. AND THE ORDER STAT. MEDIANS FROM THE LAMBDA =  1.7 DIST. IS   .99989
THE CORRELATION BETWEEN THE  3000 ORDERED OBS. AND THE ORDER STAT. MEDIANS FROM THE LAMBDA =  1.6 DIST. IS   .99985
THE CORRELATION BETWEEN THE  3000 ORDERED OBS. AND THE ORDER STAT. MEDIANS FROM THE LAMBDA =  1.5 DIST. IS   .99983
THE CORRELATION BETWEEN THE  3000 ORDERED OBS. AND THE ORDER STAT. MEDIANS FROM THE LAMBDA =  1.4 DIST. IS   .99983
THE CORRELATION BETWEEN THE  3000 ORDERED OBS. AND THE ORDER STAT. MEDIANS FROM THE LAMBDA =  1.3 DIST. IS   .99985
THE CORRELATION BETWEEN THE  3000 ORDERED OBS. AND THE ORDER STAT. MEDIANS FROM THE LAMBDA =  1.2 DIST. IS   .99989
THE CORRELATION BETWEEN THE  3000 ORDERED OBS. AND THE ORDER STAT. MEDIANS FROM THE LAMBDA =  1.1 DIST. IS   .99994
THE CORRELATION BETWEEN THE  3000 ORDERED OBS. AND THE ORDER STAT. MEDIANS FROM THE LAMBDA =  1.0 DIST. IS   .99997
THE CORRELATION BETWEEN THE  3000 ORDERED OBS. AND THE ORDER STAT. MEDIANS FROM THE LAMBDA =   .9 DIST. IS   .99994
THE CORRELATION BETWEEN THE  3000 ORDERED OBS. AND THE ORDER STAT. MEDIANS FROM THE LAMBDA =   .8 DIST. IS   .99978
THE CORRELATION BETWEEN THE  3000 ORDERED OBS. AND THE ORDER STAT. MEDIANS FROM THE LAMBDA =   .7 DIST. IS   .99938
THE CORRELATION BETWEEN THE  3000 ORDERED OBS. AND THE ORDER STAT. MEDIANS FROM THE LAMBDA =   .6 DIST. IS   .99857
THE CORRELATION BETWEEN THE  3000 ORDERED OBS. AND THE ORDER STAT. MEDIANS FROM THE LAMBDA =   .5 DIST. IS   .99710
THE CORRELATION BETWEEN THE  3000 ORDERED OBS. AND THE ORDER STAT. MEDIANS FROM THE LAMBDA =   .4 DIST. IS   .99457
THE CORRELATION BETWEEN THE  3000 ORDERED OBS. AND THE ORDER STAT. MEDIANS FROM THE LAMBDA =   .3 DIST. IS   .99040
THE CORRELATION BETWEEN THE  3000 ORDERED OBS. AND THE ORDER STAT. MEDIANS FROM THE LAMBDA =   .2 DIST. IS   .98368
THE CORRELATION BETWEEN THE  3000 ORDERED OBS. AND THE ORDER STAT. MEDIANS FROM THE NORMAL DISTRIBUTION IS   .97770
THE CORRELATION BETWEEN THE  3000 ORDERED OBS. AND THE ORDER STAT. MEDIANS FROM THE LAMBDA =   .1 DIST. IS   .97302
THE CORRELATION BETWEEN THE  3000 ORDERED OBS. AND THE ORDER STAT. MEDIANS FROM THE LOGISTIC DIST.       IS   .95625
THE CORRELATION BETWEEN THE  3000 ORDERED OBS. AND THE ORDER STAT. MEDIANS FROM THE DOUBLE EXP. DIST.    IS   .92059
THE CORRELATION BETWEEN THE  3000 ORDERED OBS. AND THE ORDER STAT. MEDIANS FROM THE LAMBDA =  -.1 DIST. IS   .93013
THE CORRELATION BETWEEN THE  3000 ORDERED OBS. AND THE ORDER STAT. MEDIANS FROM THE LAMBDA =  -.2 DIST. IS   .89000
THE CORRELATION BETWEEN THE  3000 ORDERED OBS. AND THE ORDER STAT. MEDIANS FROM THE LAMBDA =  -.3 DIST. IS   .83028
THE CORRELATION BETWEEN THE  3000 ORDERED OBS. AND THE ORDER STAT. MEDIANS FROM THE LAMBDA =  -.4 DIST. IS   .74672
THE CORRELATION BETWEEN THE  3000 ORDERED OBS. AND THE ORDER STAT. MEDIANS FROM THE LAMBDA =  -.5 DIST. IS   .64101
THE CORRELATION BETWEEN THE  3000 ORDERED OBS. AND THE ORDER STAT. MEDIANS FROM THE LAMBDA =  -.6 DIST. IS   .52412
THE CORRELATION BETWEEN THE  3000 ORDERED OBS. AND THE ORDER STAT. MEDIANS FROM THE LAMBDA =  -.7 DIST. IS   .41218
THE CORRELATION BETWEEN THE  3000 ORDERED OBS. AND THE ORDER STAT. MEDIANS FROM THE LAMBDA =  -.8 DIST. IS   .31760
THE CORRELATION BETWEEN THE  3000 ORDERED OBS. AND THE ORDER STAT. MEDIANS FROM THE LAMBDA =  -.9 DIST. IS   .24457
THE CORRELATION BETWEEN THE  3000 ORDERED OBS. AND THE ORDER STAT. MEDIANS FROM THE CAUCHY DISTRIBUTION IS   .19533
THE CORRELATION BETWEEN THE  3000 ORDERED OBS. AND THE ORDER STAT. MEDIANS FROM THE LAMBDA = -1.0 DIST. IS   .19139
THE CORRELATION BETWEEN THE  3000 ORDERED OBS. AND THE ORDER STAT. MEDIANS FROM THE LAMBDA = -1.1 DIST. IS   .15330
THE CORRELATION BETWEEN THE  3000 ORDERED OBS. AND THE ORDER STAT. MEDIANS FROM THE LAMBDA = -1.2 DIST. IS   .12657
THE CORRELATION BETWEEN THE  3000 ORDERED OBS. AND THE ORDER STAT. MEDIANS FROM THE LAMBDA = -1.3 DIST. IS   .10765
THE CORRELATION BETWEEN THE  3000 ORDERED OBS. AND THE ORDER STAT. MEDIANS FROM THE LAMBDA = -1.4 DIST. IS   .09409
THE CORRELATION BETWEEN THE  3000 ORDERED OBS. AND THE ORDER STAT. MEDIANS FROM THE LAMBDA = -1.5 DIST. IS   .08419
THE CORRELATION BETWEEN THE  3000 ORDERED OBS. AND THE ORDER STAT. MEDIANS FROM THE LAMBDA = -1.6 DIST. IS   .07683
THE CORRELATION BETWEEN THE  3000 ORDERED OBS. AND THE ORDER STAT. MEDIANS FROM THE LAMBDA = -1.7 DIST. IS   .07124
THE CORRELATION BETWEEN THE  3000 ORDERED OBS. AND THE ORDER STAT. MEDIANS FROM THE LAMBDA = -1.8 DIST. IS   .06692
THE CORRELATION BETWEEN THE  3000 ORDERED OBS. AND THE ORDER STAT. MEDIANS FROM THE LAMBDA = -1.9 DIST. IS   .06350
THE CORRELATION BETWEEN THE  3000 ORDERED OBS. AND THE ORDER STAT. MEDIANS FROM THE LAMBDA = -2.0 DIST. IS   .06076
```

Figure 23. Printout plot correlation coefficient analysis for Rand uniform random numbers

For the Josephson Junction voltage counts data (see fig. 24), the MPPCC criterion indicated that the normal distribution was the best fit which is in agreement with the already-seen linearity of the (fig. 17) normal probability plot. For the wind velocity data (see fig. 25), the MPPCC criterion indicated that the normal distribution (though not precisely optimal) was nevertheless near-optimal and is in agreement with the near-linearity of the (fig. 18) normal probability plot. The MPPCC criteria as applied to the beam deflection data is presented in figure 26. As expected, the best-fit symmetric distribution to this data set is in the U-shaped distribution region. The analysis of the x-ray crystallography residuals (see fig. 27) also confirms what was previously suspected--*viz.*, that the best distribution is in the moderate to moderate long-tailed region.

The final example (see fig. 28) illustrates the use of the MPPCC for a skewed distributional family--in this case the extreme value family. The data set considered is annual maximum wind speeds at Walla Walla, Washington. In a fashion similar to the symmetric family analysis, a representative set of 46 members of the extreme value distributional family were selected and the probability plot correlation coefficient was computed for each. The results of the analysis indicated that an extreme value distribution with shape parameter $\gamma = 7$ yields the best fit.

The use of the MPPCC criteria is recommended not as a replacement for examination of individual probability plots, but rather as an important complement to such analyses. The automated procedures presented above allow the analyst to quickly "converge" to a neighborhood of distributions which provide good fits to the data set under examination.

4-PLOT ANALYSIS

The general question posed in this section is as follows: Given that one would like to prepare a completely automated (computerized) first-pass analysis that would be applicable to a wide variety of data sets and which would take no more than one computer page, what would one include on that single page?

As has been stressed throughout this paper, the basic assumptions which must be tested to assure that a measurement process is in control include randomness, fixed location, fixed variation, and fixed distribution.

The value of the run sequence plot and the lag-1 autocorrelation plots have already been amply discussed with respect to the first three points. The probability plot has been discussed at length with respect to the last point. And so the following 4-plot analysis is presented as an automated

```
THE CORRELATION BETWEEN THE   700 ORDERED OBS. AND THE ORDER STAT. MEDIANS FROM THE LAMBDA =  2.0 DIST. IS   .95543
THE CORRELATION BETWEEN THE   700 ORDERED OBS. AND THE ORDER STAT. MEDIANS FROM THE LAMBDA =  1.9 DIST. IS   .95447
THE CORRELATION BETWEEN THE   700 ORDERED OBS. AND THE ORDER STAT. MEDIANS FROM THE LAMBDA =  1.8 DIST. IS   .95374
THE CORRELATION BETWEEN THE   700 ORDERED OBS. AND THE ORDER STAT. MEDIANS FROM THE LAMBDA =  1.7 DIST. IS   .95308
THE CORRELATION BETWEEN THE   700 ORDERED OBS. AND THE ORDER STAT. MEDIANS FROM THE LAMBDA =  1.6 DIST. IS   .95265
THE CORRELATION BETWEEN THE   700 ORDERED OBS. AND THE ORDER STAT. MEDIANS FROM THE LAMBDA =  1.5 DIST. IS   .95244
THE CORRELATION BETWEEN THE   700 ORDERED OBS. AND THE ORDER STAT. MEDIANS FROM THE LAMBDA =  1.4 DIST. IS   .95249
THE CORRELATION BETWEEN THE   700 ORDERED OBS. AND THE ORDER STAT. MEDIANS FROM THE LAMBDA =  1.3 DIST. IS   .95267
THE CORRELATION BETWEEN THE   700 ORDERED OBS. AND THE ORDER STAT. MEDIANS FROM THE LAMBDA =  1.2 DIST. IS   .95329
THE CORRELATION BETWEEN THE   700 ORDERED OBS. AND THE ORDER STAT. MEDIANS FROM THE LAMBDA =  1.1 DIST. IS   .95407
THE CORRELATION BETWEEN THE   700 ORDERED OBS. AND THE ORDER STAT. MEDIANS FROM THE LAMBDA =  1.0 DIST. IS   .95544
THE CORRELATION BETWEEN THE   700 ORDERED OBS. AND THE ORDER STAT. MEDIANS FROM THE LAMBDA =   .9 DIST. IS   .95706
THE CORRELATION BETWEEN THE   700 ORDERED OBS. AND THE ORDER STAT. MEDIANS FROM THE LAMBDA =   .8 DIST. IS   .95911
THE CORRELATION BETWEEN THE   700 ORDERED OBS. AND THE ORDER STAT. MEDIANS FROM THE LAMBDA =   .7 DIST. IS   .96161
THE CORRELATION BETWEEN THE   700 ORDERED OBS. AND THE ORDER STAT. MEDIANS FROM THE LAMBDA =   .6 DIST. IS   .96446
THE CORRELATION BETWEEN THE   700 ORDERED OBS. AND THE ORDER STAT. MEDIANS FROM THE LAMBDA =   .5 DIST. IS   .96737
THE CORRELATION BETWEEN THE   700 ORDERED OBS. AND THE ORDER STAT. MEDIANS FROM THE LAMBDA =   .4 DIST. IS   .97048
THE CORRELATION BETWEEN THE   700 ORDERED OBS. AND THE ORDER STAT. MEDIANS FROM THE LAMBDA =   .3 DIST. IS   .97315
THE CORRELATION BETWEEN THE   700 ORDERED OBS. AND THE ORDER STAT. MEDIANS FROM THE LAMBDA =   .2 DIST. IS   .97492 MAX
THE CORRELATION BETWEEN THE   700 ORDERED OBS. AND THE ORDER STAT. MEDIANS FROM THE NORMAL DISTRIBUTION IS   .97484
THE CORRELATION BETWEEN THE   700 ORDERED OBS. AND THE ORDER STAT. MEDIANS FROM THE LAMBDA =   .1 DIST. IS   .97434
THE CORRELATION BETWEEN THE   700 ORDERED OBS. AND THE ORDER STAT. MEDIANS FROM THE LOGISTIC DIST.        IS   .97045
THE CORRELATION BETWEEN THE   700 ORDERED OBS. AND THE ORDER STAT. MEDIANS FROM THE DOUBLE EXP. DIST.     IS   .95613
THE CORRELATION BETWEEN THE   700 ORDERED OBS. AND THE ORDER STAT. MEDIANS FROM THE LAMBDA =  -.1 DIST. IS   .96028
THE CORRELATION BETWEEN THE   700 ORDERED OBS. AND THE ORDER STAT. MEDIANS FROM THE LAMBDA =  -.2 DIST. IS   .94076
THE CORRELATION BETWEEN THE   700 ORDERED OBS. AND THE ORDER STAT. MEDIANS FROM THE LAMBDA =  -.3 DIST. IS   .90818
THE CORRELATION BETWEEN THE   700 ORDERED OBS. AND THE ORDER STAT. MEDIANS FROM THE LAMBDA =  -.4 DIST. IS   .85924
THE CORRELATION BETWEEN THE   700 ORDERED OBS. AND THE ORDER STAT. MEDIANS FROM THE LAMBDA =  -.5 DIST. IS   .79357
THE CORRELATION BETWEEN THE   700 ORDERED OBS. AND THE ORDER STAT. MEDIANS FROM THE LAMBDA =  -.6 DIST. IS   .71518
THE CORRELATION BETWEEN THE   700 ORDERED OBS. AND THE ORDER STAT. MEDIANS FROM THE LAMBDA =  -.7 DIST. IS   .63128
THE CORRELATION BETWEEN THE   700 ORDERED OBS. AND THE ORDER STAT. MEDIANS FROM THE LAMBDA =  -.8 DIST. IS   .55017
THE CORRELATION BETWEEN THE   700 ORDERED OBS. AND THE ORDER STAT. MEDIANS FROM THE LAMBDA =  -.9 DIST. IS   .47743
THE CORRELATION BETWEEN THE   700 ORDERED OBS. AND THE ORDER STAT. MEDIANS FROM THE CAUCHY DISTRIBUTION IS   .42313
THE CORRELATION BETWEEN THE   700 ORDERED OBS. AND THE ORDER STAT. MEDIANS FROM THE LAMBDA = -1.0 DIST. IS   .41619
THE CORRELATION BETWEEN THE   700 ORDERED OBS. AND THE ORDER STAT. MEDIANS FROM THE LAMBDA = -1.1 DIST. IS   .36553
THE CORRELATION BETWEEN THE   700 ORDERED OBS. AND THE ORDER STAT. MEDIANS FROM THE LAMBDA = -1.2 DIST. IS   .32518
THE CORRELATION BETWEEN THE   700 ORDERED OBS. AND THE ORDER STAT. MEDIANS FROM THE LAMBDA = -1.3 DIST. IS   .29315
THE CORRELATION BETWEEN THE   700 ORDERED OBS. AND THE ORDER STAT. MEDIANS FROM THE LAMBDA = -1.4 DIST. IS   .26772
THE CORRELATION BETWEEN THE   700 ORDERED OBS. AND THE ORDER STAT. MEDIANS FROM THE LAMBDA = -1.5 DIST. IS   .24749
THE CORRELATION BETWEEN THE   700 ORDERED OBS. AND THE ORDER STAT. MEDIANS FROM THE LAMBDA = -1.6 DIST. IS   .23128
THE CORRELATION BETWEEN THE   700 ORDERED OBS. AND THE ORDER STAT. MEDIANS FROM THE LAMBDA = -1.7 DIST. IS   .21820
THE CORRELATION BETWEEN THE   700 ORDERED OBS. AND THE ORDER STAT. MEDIANS FROM THE LAMBDA = -1.8 DIST. IS   .20752
THE CORRELATION BETWEEN THE   700 ORDERED OBS. AND THE ORDER STAT. MEDIANS FROM THE LAMBDA = -1.9 DIST. IS   .19871
THE CORRELATION BETWEEN THE   700 ORDERED OBS. AND THE ORDER STAT. MEDIANS FROM THE LAMBDA = -2.0 DIST. IS   .19139
```

Figure 24. Printout plot correlation coefficient analysis for Josephson Junction cryothermometry voltage counts

```
THE CORRELATION BETWEEN THE   1200 ORDERED OBS. AND THE ORDER STAT. MEDIANS FROM THE LAMBDA =  2.0 DIST. IS   .98571
THE CORRELATION BETWEEN THE   1200 ORDERED OBS. AND THE ORDER STAT. MEDIANS FROM THE LAMBDA =  1.9 DIST. IS   .98502
THE CORRELATION BETWEEN THE   1200 ORDERED OBS. AND THE ORDER STAT. MEDIANS FROM THE LAMBDA =  1.8 DIST. IS   .98442
THE CORRELATION BETWEEN THE   1200 ORDERED OBS. AND THE ORDER STAT. MEDIANS FROM THE LAMBDA =  1.7 DIST. IS   .98393
THE CORRELATION BETWEEN THE   1200 ORDERED OBS. AND THE ORDER STAT. MEDIANS FROM THE LAMBDA =  1.6 DIST. IS   .98356
THE CORRELATION BETWEEN THE   1200 ORDERED OBS. AND THE ORDER STAT. MEDIANS FROM THE LAMBDA =  1.5 DIST. IS   .98336
THE CORRELATION BETWEEN THE   1200 ORDERED OBS. AND THE ORDER STAT. MEDIANS FROM THE LAMBDA =  1.4 DIST. IS   .98335
THE CORRELATION BETWEEN THE   1200 ORDERED OBS. AND THE ORDER STAT. MEDIANS FROM THE LAMBDA =  1.3 DIST. IS   .98355
THE CORRELATION BETWEEN THE   1200 ORDERED OBS. AND THE ORDER STAT. MEDIANS FROM THE LAMBDA =  1.2 DIST. IS   .98400
THE CORRELATION BETWEEN THE   1200 ORDERED OBS. AND THE ORDER STAT. MEDIANS FROM THE LAMBDA =  1.1 DIST. IS   .98472
THE CORRELATION BETWEEN THE   1200 ORDERED OBS. AND THE ORDER STAT. MEDIANS FROM THE LAMBDA =  1.0 DIST. IS   .98571
THE CORRELATION BETWEEN THE   1200 ORDERED OBS. AND THE ORDER STAT. MEDIANS FROM THE LAMBDA =   .9 DIST. IS   .98701
THE CORRELATION BETWEEN THE   1200 ORDERED OBS. AND THE ORDER STAT. MEDIANS FROM THE LAMBDA =   .8 DIST. IS   .98858
THE CORRELATION BETWEEN THE   1200 ORDERED OBS. AND THE ORDER STAT. MEDIANS FROM THE LAMBDA =   .7 DIST. IS   .99040
THE CORRELATION BETWEEN THE   1200 ORDERED OBS. AND THE ORDER STAT. MEDIANS FROM THE LAMBDA =   .6 DIST. IS   .99240
THE CORRELATION BETWEEN THE   1200 ORDERED OBS. AND THE ORDER STAT. MEDIANS FROM THE LAMBDA =   .5 DIST. IS   .99441
THE CORRELATION BETWEEN THE   1200 ORDERED OBS. AND THE ORDER STAT. MEDIANS FROM THE LAMBDA =   .4 DIST. IS   .99620
THE CORRELATION BETWEEN THE   1200 ORDERED OBS. AND THE ORDER STAT. MEDIANS FROM THE LAMBDA =   .3 DIST. IS   .99733 MAX
THE CORRELATION BETWEEN THE   1200 ORDERED OBS. AND THE ORDER STAT. MEDIANS FROM THE LAMBDA =   .2 DIST. IS   .99709
THE CORRELATION BETWEEN THE   1200 ORDERED OBS. AND THE ORDER STAT. MEDIANS FROM THE NORMAL DISTRIBUTION IS   .99579
THE CORRELATION BETWEEN THE   1200 ORDERED OBS. AND THE ORDER STAT. MEDIANS FROM THE LAMBDA =   .1 DIST. IS   .99435
THE CORRELATION BETWEEN THE   1200 ORDERED OBS. AND THE ORDER STAT. MEDIANS FROM THE LOGISTIC DIST.       IS   .98730
THE CORRELATION BETWEEN THE   1200 ORDERED OBS. AND THE ORDER STAT. MEDIANS FROM THE DOUBLE EXP. DIST.    IS   .96786
THE CORRELATION BETWEEN THE   1200 ORDERED OBS. AND THE ORDER STAT. MEDIANS FROM THE LAMBDA =  -.1 DIST. IS   .97318
THE CORRELATION BETWEEN THE   1200 ORDERED OBS. AND THE ORDER STAT. MEDIANS FROM THE LAMBDA =  -.2 DIST. IS   .94815
THE CORRELATION BETWEEN THE   1200 ORDERED OBS. AND THE ORDER STAT. MEDIANS FROM THE LAMBDA =  -.3 DIST. IS   .90757
THE CORRELATION BETWEEN THE   1200 ORDERED OBS. AND THE ORDER STAT. MEDIANS FROM THE LAMBDA =  -.4 DIST. IS   .84763
THE CORRELATION BETWEEN THE   1200 ORDERED OBS. AND THE ORDER STAT. MEDIANS FROM THE LAMBDA =  -.5 DIST. IS   .76819
THE CORRELATION BETWEEN THE   1200 ORDERED OBS. AND THE ORDER STAT. MEDIANS FROM THE LAMBDA =  -.6 DIST. IS   .67502
THE CORRELATION BETWEEN THE   1200 ORDERED OBS. AND THE ORDER STAT. MEDIANS FROM THE LAMBDA =  -.7 DIST. IS   .57841
THE CORRELATION BETWEEN THE   1200 ORDERED OBS. AND THE ORDER STAT. MEDIANS FROM THE LAMBDA =  -.8 DIST. IS   .48857
THE CORRELATION BETWEEN THE   1200 ORDERED OBS. AND THE ORDER STAT. MEDIANS FROM THE LAMBDA =  -.9 DIST. IS   .41171
THE CORRELATION BETWEEN THE   1200 ORDERED OBS. AND THE ORDER STAT. MEDIANS FROM THE CAUCHY DISTRIBUTION IS   .35563
THE CORRELATION BETWEEN THE   1200 ORDERED OBS. AND THE ORDER STAT. MEDIANS FROM THE LAMBDA = -1.0 DIST. IS   .34955
THE CORRELATION BETWEEN THE   1200 ORDERED OBS. AND THE ORDER STAT. MEDIANS FROM THE LAMBDA = -1.1 DIST. IS   .30096
THE CORRELATION BETWEEN THE   1200 ORDERED OBS. AND THE ORDER STAT. MEDIANS FROM THE LAMBDA = -1.2 DIST. IS   .26358
THE CORRELATION BETWEEN THE   1200 ORDERED OBS. AND THE ORDER STAT. MEDIANS FROM THE LAMBDA = -1.3 DIST. IS   .23496
THE CORRELATION BETWEEN THE   1200 ORDERED OBS. AND THE ORDER STAT. MEDIANS FROM THE LAMBDA = -1.4 DIST. IS   .21298
THE CORRELATION BETWEEN THE   1200 ORDERED OBS. AND THE ORDER STAT. MEDIANS FROM THE LAMBDA = -1.5 DIST. IS   .19596
THE CORRELATION BETWEEN THE   1200 ORDERED OBS. AND THE ORDER STAT. MEDIANS FROM THE LAMBDA = -1.6 DIST. IS   .18264
THE CORRELATION BETWEEN THE   1200 ORDERED OBS. AND THE ORDER STAT. MEDIANS FROM THE LAMBDA = -1.7 DIST. IS   .17208
THE CORRELATION BETWEEN THE   1200 ORDERED OBS. AND THE ORDER STAT. MEDIANS FROM THE LAMBDA = -1.8 DIST. IS   .16361
THE CORRELATION BETWEEN THE   1200 ORDERED OBS. AND THE ORDER STAT. MEDIANS FROM THE LAMBDA = -1.9 DIST. IS   .15674
THE CORRELATION BETWEEN THE   1200 ORDERED OBS. AND THE ORDER STAT. MEDIANS FROM THE LAMBDA = -2.0 DIST. IS   .15109
```

Figure 25. Printout plot correlation coefficient analysis for wind velocities

```
THE CORRELATION BETWEEN THE   200 ORDERED OBS. AND THE ORDER STAT. MEDIANS FROM THE LAMBDA =  2.0 DIST. IS   .99255
THE CORRELATION BETWEEN THE   200 ORDERED OBS. AND THE ORDER STAT. MEDIANS FROM THE LAMBDA =  1.9 DIST. IS   .99308
THE CORRELATION BETWEEN THE   200 ORDERED OBS. AND THE ORDER STAT. MEDIANS FROM THE LAMBDA =  1.8 DIST. IS   .99351
THE CORRELATION BETWEEN THE   200 ORDERED OBS. AND THE ORDER STAT. MEDIANS FROM THE LAMBDA =  1.7 DIST. IS   .99383
THE CORRELATION BETWEEN THE   200 ORDERED OBS. AND THE ORDER STAT. MEDIANS FROM THE LAMBDA =  1.6 DIST. IS   .99405
THE CORRELATION BETWEEN THE   200 ORDERED OBS. AND THE ORDER STAT. MEDIANS FROM THE LAMBDA =  1.5 DIST. IS   .99417 MAX
THE CORRELATION BETWEEN THE   200 ORDERED OBS. AND THE ORDER STAT. MEDIANS FROM THE LAMBDA =  1.4 DIST. IS   .99416
THE CORRELATION BETWEEN THE   200 ORDERED OBS. AND THE ORDER STAT. MEDIANS FROM THE LAMBDA =  1.3 DIST. IS   .99403
THE CORRELATION BETWEEN THE   200 ORDERED OBS. AND THE ORDER STAT. MEDIANS FROM THE LAMBDA =  1.2 DIST. IS   .99374
THE CORRELATION BETWEEN THE   200 ORDERED OBS. AND THE ORDER STAT. MEDIANS FROM THE LAMBDA =  1.1 DIST. IS   .99326
THE CORRELATION BETWEEN THE   200 ORDERED OBS. AND THE ORDER STAT. MEDIANS FROM THE LAMBDA =  1.0 DIST. IS   .99255
THE CORRELATION BETWEEN THE   200 ORDERED OBS. AND THE ORDER STAT. MEDIANS FROM THE LAMBDA =   .9 DIST. IS   .99155
THE CORRELATION BETWEEN THE   200 ORDERED OBS. AND THE ORDER STAT. MEDIANS FROM THE LAMBDA =   .8 DIST. IS   .99015
THE CORRELATION BETWEEN THE   200 ORDERED OBS. AND THE ORDER STAT. MEDIANS FROM THE LAMBDA =   .7 DIST. IS   .98822
THE CORRELATION BETWEEN THE   200 ORDERED OBS. AND THE ORDER STAT. MEDIANS FROM THE LAMBDA =   .6 DIST. IS   .98558
THE CORRELATION BETWEEN THE   200 ORDERED OBS. AND THE ORDER STAT. MEDIANS FROM THE LAMBDA =   .5 DIST. IS   .98198
THE CORRELATION BETWEEN THE   200 ORDERED OBS. AND THE ORDER STAT. MEDIANS FROM THE LAMBDA =   .4 DIST. IS   .97703
THE CORRELATION BETWEEN THE   200 ORDERED OBS. AND THE ORDER STAT. MEDIANS FROM THE LAMBDA =   .3 DIST. IS   .97026
THE CORRELATION BETWEEN THE   200 ORDERED OBS. AND THE ORDER STAT. MEDIANS FROM THE LAMBDA =   .2 DIST. IS   .96097
THE CORRELATION BETWEEN THE   200 ORDERED OBS. AND THE ORDER STAT. MEDIANS FROM THE NORMAL DISTRIBUTION IS   .95408
THE CORRELATION BETWEEN THE   200 ORDERED OBS. AND THE ORDER STAT. MEDIANS FROM THE LAMBDA =   .1 DIST. IS   .94825
THE CORRELATION BETWEEN THE   200 ORDERED OBS. AND THE ORDER STAT. MEDIANS FROM THE LOGISTIC DIST.       IS   .93092
THE CORRELATION BETWEEN THE   200 ORDERED OBS. AND THE ORDER STAT. MEDIANS FROM THE DOUBLE EXP. DIST.    IS   .88949
THE CORRELATION BETWEEN THE   200 ORDERED OBS. AND THE ORDER STAT. MEDIANS FROM THE LAMBDA =  -.1 DIST. IS   .90760
THE CORRELATION BETWEEN THE   200 ORDERED OBS. AND THE ORDER STAT. MEDIANS FROM THE LAMBDA =  -.2 DIST. IS   .87677
THE CORRELATION BETWEEN THE   200 ORDERED OBS. AND THE ORDER STAT. MEDIANS FROM THE LAMBDA =  -.3 DIST. IS   .83721
THE CORRELATION BETWEEN THE   200 ORDERED OBS. AND THE ORDER STAT. MEDIANS FROM THE LAMBDA =  -.4 DIST. IS   .78846
THE CORRELATION BETWEEN THE   200 ORDERED OBS. AND THE ORDER STAT. MEDIANS FROM THE LAMBDA =  -.5 DIST. IS   .73138
THE CORRELATION BETWEEN THE   200 ORDERED OBS. AND THE ORDER STAT. MEDIANS FROM THE LAMBDA =  -.6 DIST. IS   .66839
THE CORRELATION BETWEEN THE   200 ORDERED OBS. AND THE ORDER STAT. MEDIANS FROM THE LAMBDA =  -.7 DIST. IS   .60309
THE CORRELATION BETWEEN THE   200 ORDERED OBS. AND THE ORDER STAT. MEDIANS FROM THE LAMBDA =  -.8 DIST. IS   .53933
THE CORRELATION BETWEEN THE   200 ORDERED OBS. AND THE ORDER STAT. MEDIANS FROM THE LAMBDA =  -.9 DIST. IS   .48027
THE CORRELATION BETWEEN THE   200 ORDERED OBS. AND THE ORDER STAT. MEDIANS FROM THE CAUCHY DISTRIBUTION IS   .44084
THE CORRELATION BETWEEN THE   200 ORDERED OBS. AND THE ORDER STAT. MEDIANS FROM THE LAMBDA = -1.0 DIST. IS   .42864
THE CORRELATION BETWEEN THE   200 ORDERED OBS. AND THE ORDER STAT. MEDIANS FROM THE LAMBDA = -1.1 DIST. IS   .38292
THE CORRELATION BETWEEN THE   200 ORDERED OBS. AND THE ORDER STAT. MEDIANS FROM THE LAMBDA = -1.2 DIST. IS   .34520
THE CORRELATION BETWEEN THE   200 ORDERED OBS. AND THE ORDER STAT. MEDIANS FROM THE LAMBDA = -1.3 DIST. IS   .31404
THE CORRELATION BETWEEN THE   200 ORDERED OBS. AND THE ORDER STAT. MEDIANS FROM THE LAMBDA = -1.4 DIST. IS   .28852
THE CORRELATION BETWEEN THE   200 ORDERED OBS. AND THE ORDER STAT. MEDIANS FROM THE LAMBDA = -1.5 DIST. IS   .26771
THE CORRELATION BETWEEN THE   200 ORDERED OBS. AND THE ORDER STAT. MEDIANS FROM THE LAMBDA = -1.6 DIST. IS   .25074
THE CORRELATION BETWEEN THE   200 ORDERED OBS. AND THE ORDER STAT. MEDIANS FROM THE LAMBDA = -1.7 DIST. IS   .23687
THE CORRELATION BETWEEN THE   200 ORDERED OBS. AND THE ORDER STAT. MEDIANS FROM THE LAMBDA = -1.8 DIST. IS   .22548
THE CORRELATION BETWEEN THE   200 ORDERED OBS. AND THE ORDER STAT. MEDIANS FROM THE LAMBDA = -1.9 DIST. IS   .21608
THE CORRELATION BETWEEN THE   200 ORDERED OBS. AND THE ORDER STAT. MEDIANS FROM THE LAMBDA = -2.0 DIST. IS   .20828
```

Figure 26. Printout plot correlation coefficient analysis for beam deflection

```
THE CORRELATION BETWEEN THE   2419 ORDERED OBS. AND THE ORDER STAT. MEDIANS FROM THE LAMBDA =  2.0 DIST. IS   .89619
THE CORRELATION BETWEEN THE   2419 ORDERED OBS. AND THE ORDER STAT. MEDIANS FROM THE LAMBDA =  1.9 DIST. IS   .89466
THE CORRELATION BETWEEN THE   2419 ORDERED OBS. AND THE ORDER STAT. MEDIANS FROM THE LAMBDA =  1.8 DIST. IS   .89334
THE CORRELATION BETWEEN THE   2419 ORDERED OBS. AND THE ORDER STAT. MEDIANS FROM THE LAMBDA =  1.7 DIST. IS   .89226
THE CORRELATION BETWEEN THE   2419 ORDERED OBS. AND THE ORDER STAT. MEDIANS FROM THE LAMBDA =  1.6 DIST. IS   .89146
THE CORRELATION BETWEEN THE   2419 ORDERED OBS. AND THE ORDER STAT. MEDIANS FROM THE LAMBDA =  1.5 DIST. IS   .89100
THE CORRELATION BETWEEN THE   2419 ORDERED OBS. AND THE ORDER STAT. MEDIANS FROM THE LAMBDA =  1.4 DIST. IS   .89095
THE CORRELATION BETWEEN THE   2419 ORDERED OBS. AND THE ORDER STAT. MEDIANS FROM THE LAMBDA =  1.3 DIST. IS   .89136
THE CORRELATION BETWEEN THE   2419 ORDERED OBS. AND THE ORDER STAT. MEDIANS FROM THE LAMBDA =  1.2 DIST. IS   .89231
THE CORRELATION BETWEEN THE   2419 ORDERED OBS. AND THE ORDER STAT. MEDIANS FROM THE LAMBDA =  1.1 DIST. IS   .89389
THE CORRELATION BETWEEN THE   2419 ORDERED OBS. AND THE ORDER STAT. MEDIANS FROM THE LAMBDA =  1.0 DIST. IS   .89619
THE CORRELATION BETWEEN THE   2419 ORDERED OBS. AND THE ORDER STAT. MEDIANS FROM THE LAMBDA =   .9 DIST. IS   .89932
THE CORRELATION BETWEEN THE   2419 ORDERED OBS. AND THE ORDER STAT. MEDIANS FROM THE LAMBDA =   .8 DIST. IS   .90339
THE CORRELATION BETWEEN THE   2419 ORDERED OBS. AND THE ORDER STAT. MEDIANS FROM THE LAMBDA =   .7 DIST. IS   .90854
THE CORRELATION BETWEEN THE   2419 ORDERED OBS. AND THE ORDER STAT. MEDIANS FROM THE LAMBDA =   .6 DIST. IS   .91490
THE CORRELATION BETWEEN THE   2419 ORDERED OBS. AND THE ORDER STAT. MEDIANS FROM THE LAMBDA =   .5 DIST. IS   .92259
THE CORRELATION BETWEEN THE   2419 ORDERED OBS. AND THE ORDER STAT. MEDIANS FROM THE LAMBDA =   .4 DIST. IS   .93173
THE CORRELATION BETWEEN THE   2419 ORDERED OBS. AND THE ORDER STAT. MEDIANS FROM THE LAMBDA =   .3 DIST. IS   .94233
THE CORRELATION BETWEEN THE   2419 ORDERED OBS. AND THE ORDER STAT. MEDIANS FROM THE LAMBDA =   .2 DIST. IS   .95428
THE CORRELATION BETWEEN THE   2419 ORDERED OBS. AND THE ORDER STAT. MEDIANS FROM THE NORMAL DISTRIBUTION IS   .96258
THE CORRELATION BETWEEN THE   2419 ORDERED OBS. AND THE ORDER STAT. MEDIANS FROM THE LAMBDA =   .1 DIST. IS   .96710
THE CORRELATION BETWEEN THE   2419 ORDERED OBS. AND THE ORDER STAT. MEDIANS FROM THE LOGISTIC DIST.         IS   .97965
THE CORRELATION BETWEEN THE   2419 ORDERED OBS. AND THE ORDER STAT. MEDIANS FROM THE DOUBLE EXP. DIST.      IS   .98785
THE CORRELATION BETWEEN THE   2419 ORDERED OBS. AND THE ORDER STAT. MEDIANS FROM THE LAMBDA =  -.1 DIST. IS   .98960
THE CORRELATION BETWEEN THE   2419 ORDERED OBS. AND THE ORDER STAT. MEDIANS FROM THE LAMBDA =  -.2 DIST. IS   .99265 MAX
THE CORRELATION BETWEEN THE   2419 ORDERED OBS. AND THE ORDER STAT. MEDIANS FROM THE LAMBDA =  -.3 DIST. IS   .98206
THE CORRELATION BETWEEN THE   2419 ORDERED OBS. AND THE ORDER STAT. MEDIANS FROM THE LAMBDA =  -.4 DIST. IS   .94981
THE CORRELATION BETWEEN THE   2419 ORDERED OBS. AND THE ORDER STAT. MEDIANS FROM THE LAMBDA =  -.5 DIST. IS   .89075
THE CORRELATION BETWEEN THE   2419 ORDERED OBS. AND THE ORDER STAT. MEDIANS FROM THE LAMBDA =  -.6 DIST. IS   .80825
THE CORRELATION BETWEEN THE   2419 ORDERED OBS. AND THE ORDER STAT. MEDIANS FROM THE LAMBDA =  -.7 DIST. IS   .71434
THE CORRELATION BETWEEN THE   2419 ORDERED OBS. AND THE ORDER STAT. MEDIANS FROM THE LAMBDA =  -.8 DIST. IS   .62288
THE CORRELATION BETWEEN THE   2419 ORDERED OBS. AND THE ORDER STAT. MEDIANS FROM THE LAMBDA =  -.9 DIST. IS   .54283
THE CORRELATION BETWEEN THE   2419 ORDERED OBS. AND THE ORDER STAT. MEDIANS FROM THE CAUCHY DISTRIBUTION IS   .48098
THE CORRELATION BETWEEN THE   2419 ORDERED OBS. AND THE ORDER STAT. MEDIANS FROM THE LAMBDA = -1.0 DIST. IS   .47729
THE CORRELATION BETWEEN THE   2419 ORDERED OBS. AND THE ORDER STAT. MEDIANS FROM THE LAMBDA = -1.1 DIST. IS   .42510
THE CORRELATION BETWEEN THE   2419 ORDERED OBS. AND THE ORDER STAT. MEDIANS FROM THE LAMBDA = -1.2 DIST. IS   .38433
THE CORRELATION BETWEEN THE   2419 ORDERED OBS. AND THE ORDER STAT. MEDIANS FROM THE LAMBDA = -1.3 DIST. IS   .35243
THE CORRELATION BETWEEN THE   2419 ORDERED OBS. AND THE ORDER STAT. MEDIANS FROM THE LAMBDA = -1.4 DIST. IS   .32730
THE CORRELATION BETWEEN THE   2419 ORDERED OBS. AND THE ORDER STAT. MEDIANS FROM THE LAMBDA = -1.5 DIST. IS   .30730
THE CORRELATION BETWEEN THE   2419 ORDERED OBS. AND THE ORDER STAT. MEDIANS FROM THE LAMBDA = -1.6 DIST. IS   .29118
THE CORRELATION BETWEEN THE   2419 ORDERED OBS. AND THE ORDER STAT. MEDIANS FROM THE LAMBDA = -1.7 DIST. IS   .27805
THE CORRELATION BETWEEN THE   2419 ORDERED OBS. AND THE ORDER STAT. MEDIANS FROM THE LAMBDA = -1.8 DIST. IS   .26722
THE CORRELATION BETWEEN THE   2419 ORDERED OBS. AND THE ORDER STAT. MEDIANS FROM THE LAMBDA = -1.9 DIST. IS   .25820
THE CORRELATION BETWEEN THE   2419 ORDERED OBS. AND THE ORDER STAT. MEDIANS FROM THE LAMBDA = -2.0 DIST. IS   .25061
```

Figure 27. Printout plot correlation coefficient analysis for x-ray crystallography residuals

CALL EXTREM(X,53)

WALLA WALLA, WASHINGTON
ANNUAL MAXIMUM WIND SPEEDS

EXTREME VALUE ANALYSIS

THE SAMPLE SIZE N = 3
THE SAMPLE MEAN = 34.6 81
THE SAMPLE STANDARD DEVIA ION = 7.0998

γ	Probability Plot Correlation Coefficient	$\hat{\mu}$	$\hat{\sigma}$	τ
1.00	.79[illegible]	32.38533	.49409	9.1775
2.00	.9473[illegible]	26.87591	4.74901	2.3773
3.00	.960[illegible]	20.89503	10.44045	1.4354
4.00	.[illegible]5	14.97999	16.29362	1.0999
5.00	.99294	9.12943	22.13881	.9330
6.00	.99376	3.32032	27.95807	.8343
7.00	.99380 MAX	-2.46220	33.75460	.7694
8.00	.99354	-8.22703	39.53359	.7237
9.00	.99317	-13.97957	45.29930	.6897
10.00	.99277	-19.72330	51.05490	.6636
11.00	.99237	-25.46048	56.80268	.6428
12.00	.99199	-31.19269	62.54431	.6260
13.00	.99164	-36.92104	68.28105	.6120
14.00	.99131	-42.64632	74.01383	.6003
15.00	.99101	-48.36914	79.74337	.5903
16.00	.99073	-54.08994	85.47023	.5816
17.00	.99048	-59.80908	91.19484	.5741
18.00	.99024	-65.52682	96.91755	.5674
19.00	.99003	-71.24338	102.63864	.5616
20.00	.98983	-76.95895	108.35835	.5563
21.00	.98964	-62.67365	114.07685	.5516
22.00	.98947	-88.38760	119.79430	.5474
23.00	.98931	-94.10090	125.51083	.5435
24.00	.98917	-99.81363	131.22655	.5400
25.00	.98903	-105.52587	136.94157	.5368
30.00	.98845	-134.08124	165.50829	.5241
35.00	.98802	-162.62997	194.06537	.5152
40.00	.98769	-191.17455	222.61634	.5086
45.00	.98743	-219.71634	251.16318	.5035
50.00	.98721	-248.25616	279.70710	.4995
60.00	.98689	-305.33195	336.78910	.4935
70.00	.98665	-362.40438	393.86602	.4893
80.00	.98647	-419.47474	450.93977	.4861
90.00	.98632	-476.54367	508.01137	.4837
100.00	.98621	-533.61156	565.08142	.4817
150.00	.98586	-818.94344	850.41982	.4759
200.00	.98568	**********	1135.74908	.4730
250.00	.98557	**********	1421.07478	.4713
300.00	.98550	**********	1706.39926	.4702
350.00	.98545	**********	1991.72139	.4694
400.00	.98541	**********	2277.04306	.4688
450.00	.98538	**********	2562.36566	.4683
500.00	.98535	**********	2847.68616	.4679
750.00	.98528	**********	4274.28717	.4668
1000.00	.98524	**********	5700.88733	.4662
∞	.98513	31.48949	5.70640	1.5619

Extreme Value Type I: $G(p) = \mu + \sigma[-\log(-\log(p))]$

Extreme Value Type II: $G(p) = \mu + \sigma[(-\log(p))^{-\gamma}]$

Figure 28. Probability plot correlation coefficient analysis for Walla Walla, Washington annual maximum wind speeds

comprehensive first-pass tool for data analysis. The four plots are as follows:

1. run sequence plot

2. lag-1 autocorrelation plot

3. histogram

4. normal probability plot

The normal probability plot was chosen because the normality assumption is most commonly employed. The histogram is included as an additional graphical point of reference in case the normality assumption is untenable. In addition, a selected set of about a dozen useful locations, variation and autocorrelation summarizing statistics is included on this single page.

This particular 4-plot analysis has proved to be invariably informative in terms of assessing the validity of the underlying assumptions in a measurement process. The analysis can be applied to both raw response data and to residuals after a multi-factor (e.g., regression, ANOVA) fit. The technique is recommended only as a first pass in a data analysis and should be complemented by more detailed analysis.

The application of this technique to several examples is now discussed. The first example (fig. 29) is the 500 Rand (24) normal random numbers of figures 15 and 22. The run sequence plot indicates fixed location and variation. The lag-1 autocorrelation plot indicates randomness. The histogram indicates a bell-shaped symmetric distribution. The normal probability plot indicates normality.

The second example (fig. 30) is the 700 Josephson Junction cryothermometry voltage counts of figures 17 and 24. The run sequence plot indicates fixed location and variation and also the rather discrete nature of the data. The lag-1 autocorrelation plot indicates randomness and reinforces the discrete aspect of the data; the histogram indicates symmetry and a bell-shape; the normal probability plot indicates normality.

The third example (fig. 31) is the 200 beam deflections data on figures 19 and 26. The run sequence plot indicates fixed location and variation and perhaps a single outlier (high). The lag-1 autocorrelation plot indicates well-defined nonrandomness and additional evidence for an outlier. The histogram indicates symmetry and a U-shaped indicates that a distribution shorter-tailed than normal is needed) and additional outlier evidence. A remodeling to take into account the dominant autocorrelation structure of the data is clearly called for in this case.

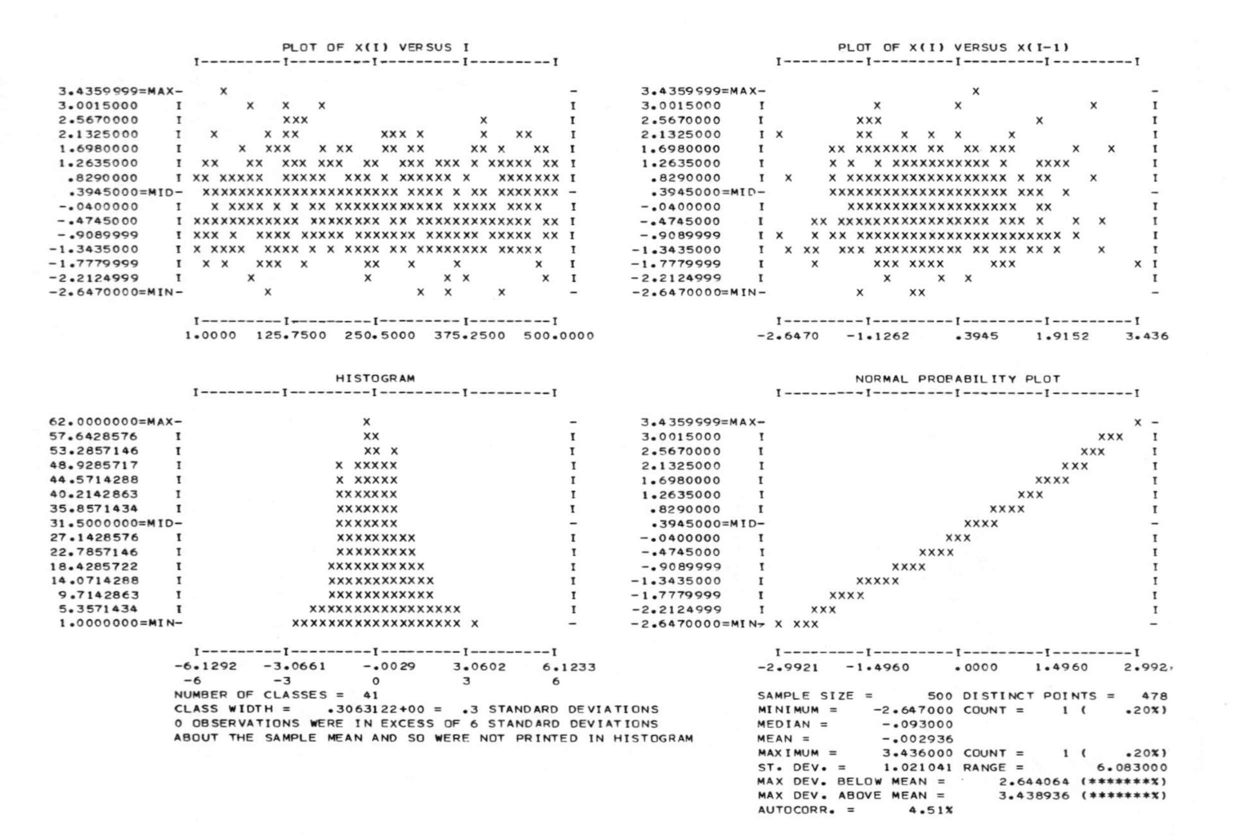

Figure 29. 4-plot analysis of Rand normal random deviates

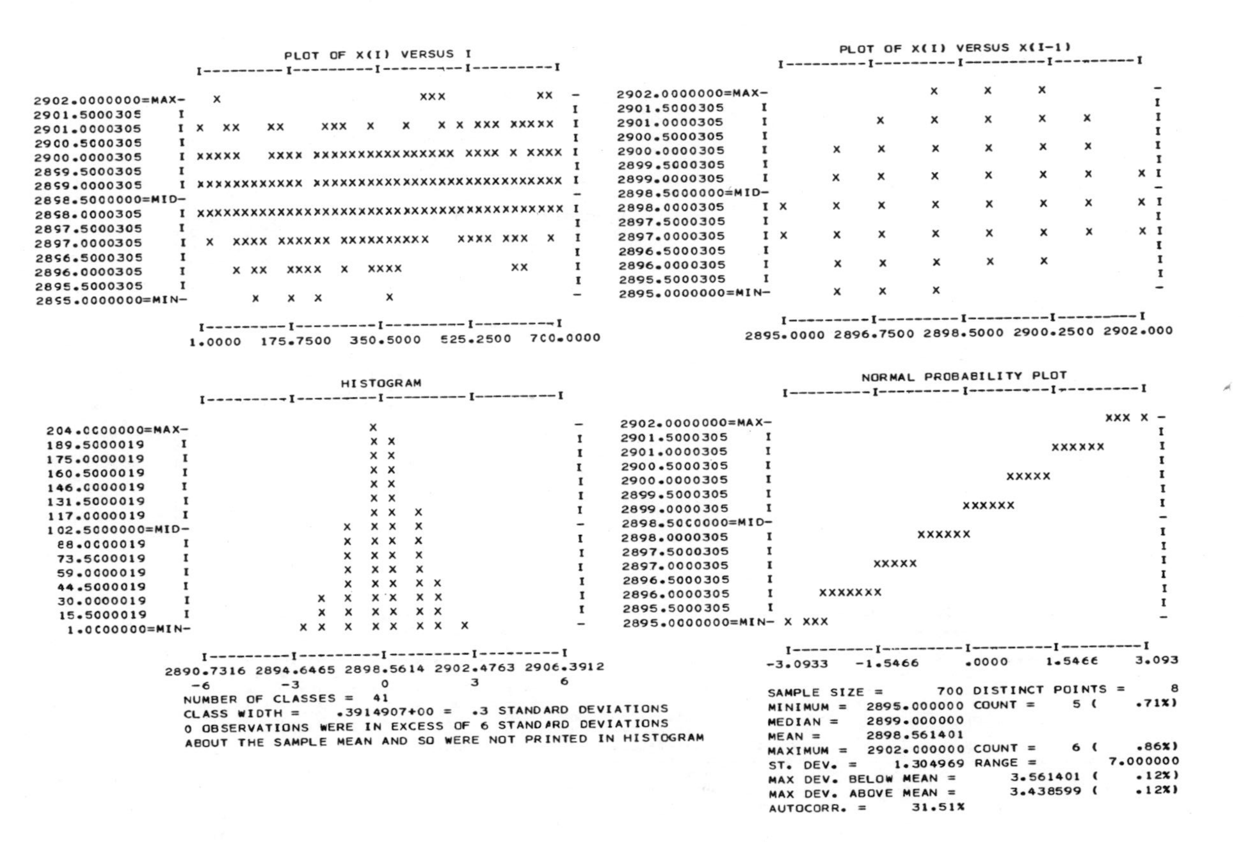

Figure 30. 4-plot analysis of Josephson Junction cryothermometry voltage counts

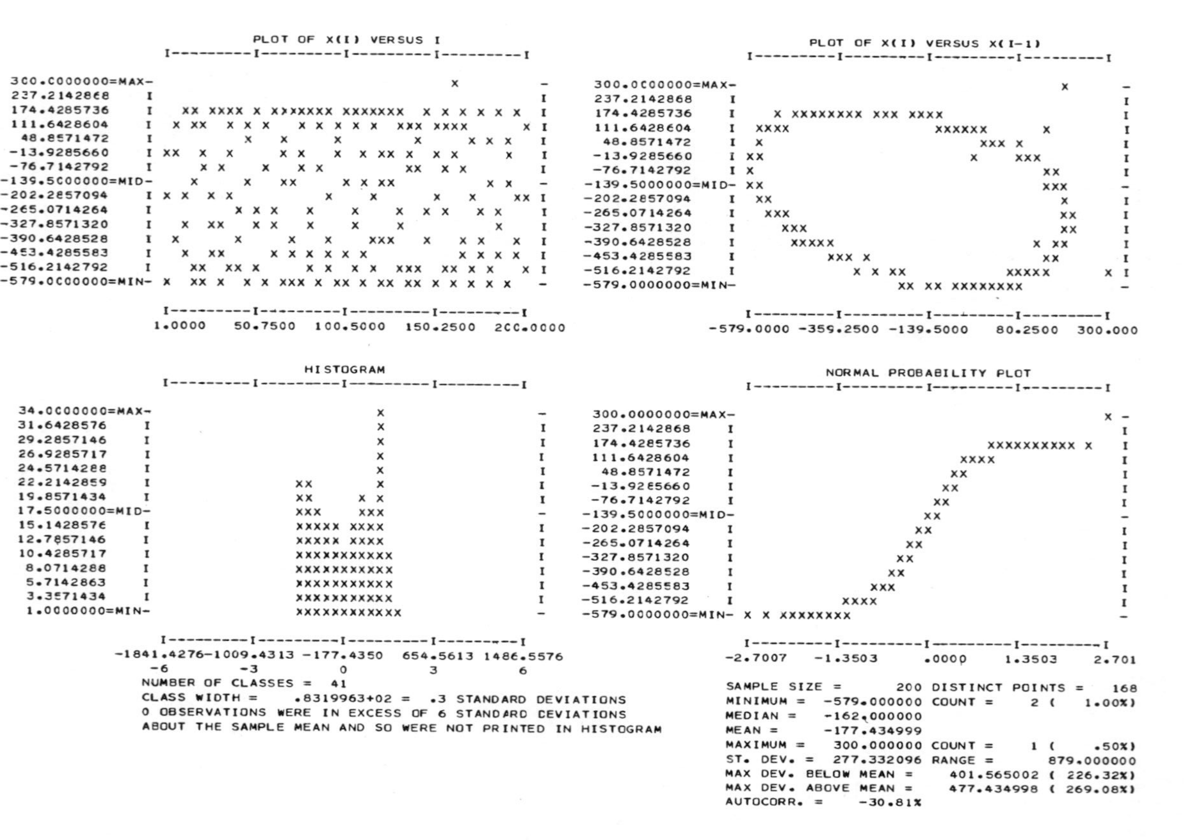

Figure 31. 4-plot analysis of beam deflection data

The fourth example (fig. 32) is the residuals from the x-ray crystallography fit. The run sequence plot indicates fixed location but larger variation (or perhaps a few high outliers) at the beginning of the set. The lag-1 autocorrelation plot indicates randomness but additional evidence for the existence of outliers. The histogram indicates symmetry and bell-shapedness but also the suggestion of longer-than-normal tails. The normal probability plot indicates non-normality--the N-shape implies that a distribution longer-tailer than normal is needed. The elongated upper right tail of the normal probability plot also gives further corroboration to the possible existence of outliers as seen initially in the run sequence plot.

The last example (fig. 33) is the 50 spectrophotometric measurements of transmittance analyzed in figures 3 and 4 of section 4. The run sequence plot indicates a shift in location in the second half of the data. The histogram is rather nondescript aside from its definite biomodal character. The normal probability plot indicates non-normality, and a shorter-tailed distribution is needed. A remodeling to take into account the dominant autocorrelation structure of the data is clearly needed.

CONCLUSION

In any given measurement process, the ultimate concern is predictability--being able to make probability statements about future output from the process. To achieve such predictability, a measurement process must be "in control." This will be the case when the output from the process behaves like random drawings from some fixed distribution with fixed location and fixed variation. The core of this paper has been to discuss various techniques to test the four components (random, fixed location, fixed variation, and fixed distribution) in the above definition of "in control." These four components are implicitly assumed in the analysis of all measurement processes.

The randomness assumption helps assure that the experimentalist does in fact have as many <u>independent</u> realizations of the phenomenon of interest as is believed. If the randomness assumption is untenable, this is frequently indicative of either the existence of some extraneous variable (which has not yet been accounted for) or alternatively (at times) an overly-fast sampling rate.

The fixed location and fixed variation assumptions help assure that the process is <u>stable</u> in the simplest sense. A process which is drifting in either its typical value (location) or its typical spread (variation) cannot be considered as "in control" and certainly negates any possibility of generating probability (predictability) statements about future output from

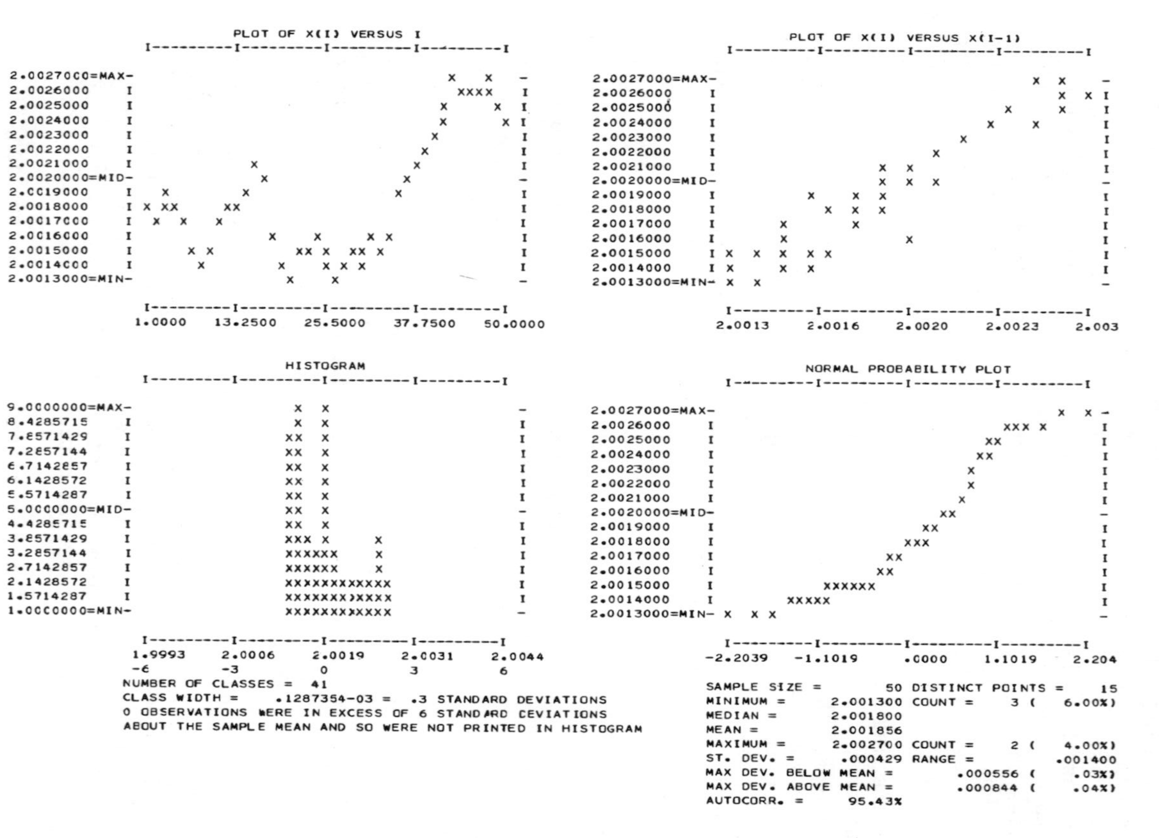

Figure 32. 4-plot analysis of x-ray crystallography residuals

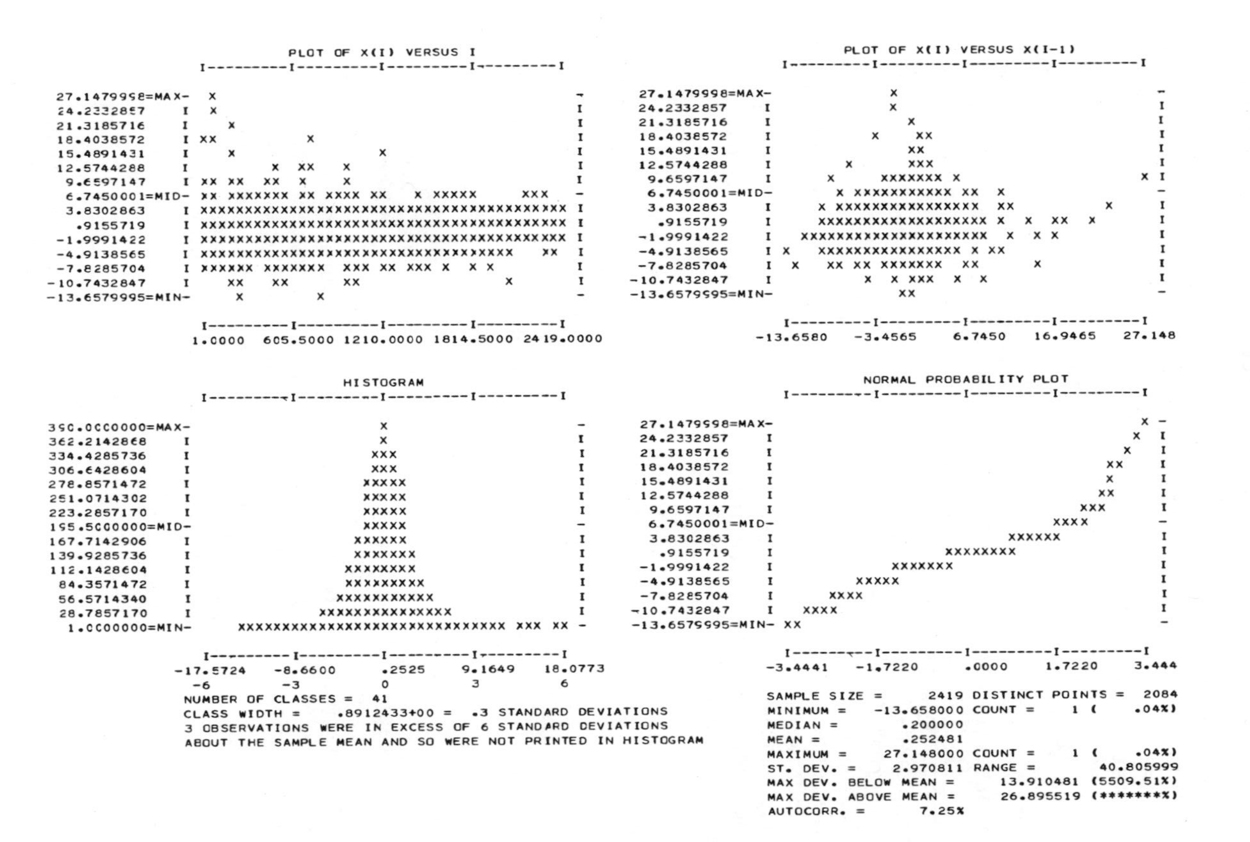

Figure 33. 4-plot analysis of spectrophotometric measurements of transmittance

the process. Problems associated with drifting in either location or variation are generally traceable--from an analysis point of view--to the existence of an unaccounted variable which is influencing the response. From an experimentalist's point of view, this variable may be (to name just a few) instrumental, procedural, or environmental in nature and almost always necessitates a scrutinous, detailed review of the experimental process as a whole. References (1), (2) and (13) are relevant in this regard.

The fixed distribution assumption is particularly important. In many (but not all) physical science experiments, the normal distribution is an appropriate distributional model. If normality is theoretically appropriate, and if the distributional tests as described in sections 10 and 11 confirm such normality, this gives considerable support to the conclusion that the measurement process is indeed "in control." On the other hand, if normality is appropriate and yet the distributional tests indicate non-normality, then this is usually another indicator that the response is being influenced by some other nonrandom variable which the analyst has not yet taken into account.

Two distinct and opposite situations may be identified with respect to this assumption. First, the nature of the measurement process may be such that there exists an *a priori* theoretical basis which dictates what distribution the output from the process should follow; and second, *no* such *a priori* theory exists for predicting what distribution the output from the process will follow.

In the first case, the normal distribution is often (but not universally) theoretically appropriate; other commonly-occurring distributional models are the Poisson distribution (*e.g.* for counting processes) and the Weibull/extreme value distributions (*e.g.*, for lifetimes of fracture processes). If a certain type of distribution is theoretically appropriate, and if the distributional tests as described in sections 10 and 11 confirm such a distribution, then this gives considerable support to the conclusion that the measurement process is indeed "in control." On the other hand, if the tests indicate a best-fit distribution that is different from theory, this is usually another indication that the response is being influenced by some other variable which the analyst has not yet taken into account.

In the second case (when the theoretical distribution is unknown), the analyst has less definitive information upon which to act. Practically speaking, in such a case, a best-fit distribution to normal is frequently an indication of stability in the measurement process. On the other hand, a best-fit distribution which is very long-tailed (*e.g*, one in the vicinity of the Cauchy distribution) is almost invariably an indication of a process which still has unresolved sources of variation.

In any event, for both of the above cases (known versus unknown theoretical distributions), it is seen that one of the obvious pieces of information which must be reported in the summary description of any measurement process is the nature of the best-fit distribution. The importance of providing such distributional information is reaffirmed when it is recalled that the ultimate objective in a measurement process is predicability and in this regard the analyst is constantly brought back to the fact that it is the underlying distribution (either known or estimated) which must be employed to form such prediction (probability) statements.

With respect to the various techniques themselves for testing randomness, location, variation, and distributional aspects of the data, it is noted that most were graphical in nature. The techniques have wide applicability due to the fact that they can be applied not only to the raw data from univariate models, but also to the residuals after the fit in multi-factor models (e.g., regression, ANOVA). With the exception of the graphical ANOVA technique, the procedures covered have been implemented as stand-alone subroutines in the portable DATAPAC FORTRAN (6,7) data analysis package which is available from the author.

ACKNOWLEDGMENT

The author is indebted to the many NBS scientists whose data have served as the basis for the various examples used in this paper. It is to be noted that the examples have been chosen to illustrate specific statistical techniques and to emphasize particular statistical anomalies. Such anomalies most frequently are uncovered in preliminary investigation of a measurement process as was the case in many of the examples cited.

The author is also pleased to acknowledge the valuable comments offered by Joseph Cameron, Joan Rosenblatt, James DeVoe and Judy Gilsinn--all members of the NBS Technical Staff.

ABSTRACT

This paper concerns itself with the important problem of testing the validity of the basic assumptions in a measurement process. The paper covers four principal areas in this regard: 1) the assumptions that are typically made in a measurement process, 2) the consequences to conclusions drawn from a measurement process if the assumptions do not hold, 3) theoretical statistical tests for the checking of basic assumptions, and 4) practical tools to facilitate the checking of basic assumptions. Examples of assumption-checking on data drawn from the chemical and physical sciences are included.

Literature Cited

1. Cameron, Joseph M., Measurement Assurance, Journal of Quality Technology, (1976), Vol. 8, No. 1, p 53.
2. Cameron, Joseph M., Procedures for the Assurance of the Adequacy of Sound Level Measurements. National Bureau of Standards Technical Note 931. Chapter 4 of Environmental Effects on Microphones and Type 2 Sound Level Meters (edited by Edward B. Magrab), (1976), p 63.
3. Eisenhart, C., Realistic Evaluation of the Precision and Accuracy of Instrument Calibration Systems. Journal of Research of the National Bureau of Standards-C. Engineering and Instrumentation, (1963), Vol. 67C, No. 2, p 161.
4. Cleveland, William S. and Kleiner, Beat., A Graphical Technique of Enhancing Scatterplots With Moving Statistics. Technometrics, (1975), Vol. 17, No. 4, p 447.
5. Currie, Lloyd A., Filliben, James J. and DeVoe, James R., Statistical and Mathematical Methods in Analytical Chemistry., Analytical Chemistry Reviews, (1972), Vol. 44, No. 5, p 479R.
6. Filliben, James J., DATAPAC: A Data Analysis Package. Proceedings of the Ninth Interface on Computer Science and Statistics, Prindle, Weber and Schmidt, Inc., Boston, (1977), p 212.
7. Filliben, James J., A User's Guide to the DATAPAC Data Analysis Package. National Bureau of Standards Technical Note (in preparation) (1977).
8. Himmelblau, David M., Process Analysis by Statistical Methods, John Wiley, New York, (1968), p 78.
9. Levene, H. and Wolfowitz, J., The Convariance Matrix and Runs Up and Down, The Annals of Mathematical Statistics, (1944), Vol. 15, p 58.
10. Hamaker, H. C., New Techniques of Statistical Teaching. Review of the International Statistical Institute, (1971), Vol. 39, No. 3, p 351.
11. Filliben, James J., User's Guide to DATAPLOT: A System for Interactive Graphical Analysis. National Bureau of Standards Technical Note (in preparation) (1977).
12. Youden, W. J., Graphical Analysis of Interlaboratory Test Results. Industrial Quality Control, (1959), Vol. 15, No. 11, p 24. (Reprinted in Journal of Quality Technology, (1972), Vol. 4, No. 1, p 29.
13. Youden, W. J., Experimental Design and ASTM Committees. Materials Research and Standards, (1961), Vol. 1, No. 11, p 862.

14. Filliben, James J., Techniques for Tail Length Analysis. Proceedings of the 18th Conference on the Design of Experiments in Army Research Development and Testing, October (1973), Part 2, ARO Report 73-2, p 425.
15. Filliben, James J., The Probability Plot Correlation Coefficient Test for Normality. Technometrics, (1975), Vol. 17, No. 1, p 111.
16. Ryan, Thomas A., Jr. and Joiner, Brian L., Normal Probability Plots and Tests for Normality, (1975), Pennsylvania State University Report.
17. Daniel Cuthbert, Use of Half-Normal Plots in Interpreting Factorial Two-Level Experiments. Technometrics, (1959), Vol. 1, No. 4, p 311.
18. Hahn, G. and Shapiro, S., Statistical Methods in Engineering, John Wiley, New York, (1967), p 260.
19. Mood, Alexander M. and Grayhill, Franklin A., Introduction to the Theory of Statistics, McGraw-Hill, New York (1963).
20. Nelson, W. and Thompson, V. C., Weibull Probability Papers. Journal of Quality Technology, (1971), Vol. 3, No. 2, p 45.
21. Wilk, M. and Gnanadesikan, R., Probability Plotting Methods for the Analysis Data, Biometrika, (1968), Vol. 55, No. 1, p 1.
22. Johnson, Norman L. and Kotz, Samuel, Continuous Univariate Distributions-1 and 2, Houghton Mifflin Company, Boston (1970).
23. Johnson, Norman L. and Kotz, Samuel, Discrete Distributions, Houghton Mifflin Company, Boston (1969).
24. Rand Corporation, A Million Random Digits With 100,000 Normal Deviates, The Free Press, Glencoe, Illinois (1955).
25. Filliben, James J., Simple and Robust Linear Estimation of the Location Parameter of a Symmetric Distribution. Unpublished Ph.D. Dissertation, Princeton University (1969).
26. Joiner, B. L. and Rosenblatt, J. R., Some Properties of the Range in Samples from Tukey's Symmetric Lambda Distributions. Journal of the American Statistical Association, (1971), Vol. 66, p 394.
27. Simiu, Emil and Filliben, James J., Probability Distributions of Extreme Wind Speeds, Journal of the Structural Division, American Society of Civil Engineers, (1976), Vol. 102, No. ST9, p 1861.
28. Simiu, Emil and Filliben, James J., Statistical Analysis of Extreme Winds, NBS Technical Note 868, (1975), p 1.

3

Systematic Error in Chemical Analysis

L. A. CURRIE and J. R. DEVOE

Analytical Chemistry Division, Institute for Materials Research, National Bureau of Standards, Washington, DC 20234

The fundamental limitation to accuracy in chemical analysis is systematic error. Unfortunately, systematic error--which comprises all nonrandom deviations of analytical results from the truth--is the rule in analytical chemistry. Systematic error comes about whenever the actual nature of the analytical process differs from that assumed. It results from invalid sampling, operator or equipment instability and blunders, unrecognized sample loss or contamination, poor instrument calibration, inadequate physical (mathematical) or random error distribution models, and faulty reporting of data. These problems, which will be covered in some detail below, are not exceptional. It is only through exhaustive, quantitative evaluation of the individual and collective effects of such violations in assumption that the analyst can hope to provide meaningful bounds for systematic error.

The impact of erroneous analytical measurements can be considerable. A recent New York Times article (1) entitled, "Medical Labs May Not Be All That Accurate" pointed up the fact that in a survey of the clinical laboratories involved in Interstate Commerce (and consequently under the monitoring of the Federal Center for Disease Control, USPHS) 31 percent were unable to identify sickle-cell anemia from blood smears. Additional tests such as hemoglobin and electrolyte content in blood were unsatisfactory in a similar fraction of laboratories. Naturally, this situation has resulted in some lack of confidence on the part of the physician; and confidence-erosion can be dangerous. Instances have occurred where a test result deviated from the norm to such an extent that the physician who ignored the result (assuming laboratory error, when there was none) made an improper diagnosis with serious consequences to the patient.

Another example of somewhat less immediate severity but greater long term importance is the measurement of ozone in the atmosphere. Figure 1 shows the deviations from the true concentration (2) of experimental results from a number of laboratories. Collectively, one sees that the laboratories produced results whose (negative) bias exceeds the imprecision bound. As a result of both systematic and random error components reported O_3 concentrations were too low by 20 percent to 60 percent (at the Air Quality Standard level). In this case the "true" concentration was provided to the testing laboratories in the form of an accurately-prepared gaseous reference sample. This example raises an important point related to the role of reference materials for transferring accuracy from one laboratory to another. Though reference materials are exceedingly useful for disclosing laboratory error, they do not eliminate the need for the quantitative assessment of all potential sources of bias.

The overwhelming importance of the systematic component of error may be grasped from eq. (1):

total error: $$e = \delta + \Delta \quad (1a)$$

random error: $$\delta = z \cdot SE = z(\sigma/\sqrt{n}) \quad (1b)$$

Where e represents the total error in $\bar{x}$ and δ and Δ represent the random and systematic components, respectively.** If normality is assumed (Gaussian random error distribution), the random error is simply the product of the random normal deviate (z) and the standard error (SE). The standard error in $\bar{x}$ depends upon the precision parameter σ (standard deviation) and the number of replications n. With increased replication the standard error tends toward zero, with the result that the total error asymptotically approaches the bias--*i.e.*, $e \rightarrow \Delta$. The ultimate capability of any analytical procedure thus rests upon the magnitude of the bias. The problem is compounded by the fact that only the precision may be directly estimated through experiment (replication). The two examples cited above simply illustrate the consequences of ignoring or giving inadequate attention to this extremely important, but more difficult to estimate, systematic component of error.

When adequate care is given to estimating bounds for Δ, the results may appear surprising. For example, in the most recent tabulation of Eu-152 γ-ray decay probabilities, the estimated limits for systematic error exceed the standard error by factors of 2.5 to 40 (3). As stated bounds for systematic error are highly dependent upon the scientific judgment and philosophy of the experimenter, under- and over-estimation of such bounds can completely cloud the meaning of analytical results. An incisive discussion of this particular problem, as related to the fundamental physical constants, has been given by Taylor *et al.* (4).

**A list of terms and symbols is given at the end of the Chapter.

In the discussion which follows we shall first examine the means and limitations of nonrandom error detection. A systematic analysis of the individual steps of the Chemical Measurement Process (CMP) will then be undertaken in order to expose the sources and methods for controlling this component of error. Finally, some simple, yet powerful diagnostic techniques will be presented for the identification of bias and blunders affecting experimental results.

SYSTEMATIC ERROR BOUNDS

Limits for systematic error may be arrived at in two different ways. (1) They may be estimated (in the statistical sense) by comparing an experimental result $\bar{x}$ with the true value τ (if known) or with values obtained by <u>independent</u>, <u>reliable</u> methods or laboratories.*** As true values in the strictest sense exist only by definition, the first type of comparison generally implies the availability of "accepted" or "certified" values, such as those which accompany reference materials distributed by national standardizing laboratories. (2) The second approach to systematic error evaluation is through detailed analysis of the structure of the CMP in order to infer bounds for overall propagated systematic error. This approach, in contrast to the former, relies wholly upon sound, scientific judgment. The first, empirical approach thus yields

$$\hat{\Delta} = e = \bar{x} - \tau \tag{2a}$$

for the estimated bias, and

$$\hat{\Delta}_{\pm} = e \pm \delta_M \tag{2b}$$

for its upper and lower limits. In eq. (2b), δ_M represents the absolute value of the maximum likely random error--commonly taken to be two to three times the standard error. The second, "theoretical" approach yields inferred bounds

$$\tilde{\Delta}_{\pm} = P(\tilde{\Delta}_{\pm})_i \tag{3}$$

where $(\tilde{\Delta}_{+})_i$ represents the contribution of step-i of the CMP, and P symbolizes the appropriate propagation operation which in the simplest case is merely summation--<u>e.g.</u>, $\tilde{\Delta}_{\pm} = \Sigma\,(\tilde{\Delta}_{\pm})_i$.

It is essential that <u>both</u> types of analysis take place. Verification of measurement accuracy can only come through intercomparison. ($\hat{\Delta}_{\pm}$, method-1 must cover zero for an unbiased

***Intercomparisons lacking either independence or reliability are fruitless. A dramatic illustration has been given by Yolken <u>(5)</u>, who contrasts results obtained by expert laboratories with those obtained using "certification by concensus" of nonexperts.

measurement process.) However, meaningful uncertainty bounds for any given experiment ultimately depend upon careful evaluation of $\tilde{\Delta}_{\pm}$ (method-2). Furthermore, having both types of estimate makes possible an extremely valuable check for consistency: the overlap of $\hat{\Delta}_{\pm}$ and $\tilde{\Delta}_{\pm}$. (Further discussion of bounds for systematic and total error will be given in the section on reporting of data.)

Although scientific evaluation of systematic error bounds ($\tilde{\Delta}_{\pm}$) is quite difficult, adequate estimation <u>via</u> intercomparison ($\hat{\Delta}_{\pm}$) is perhaps even more difficult. This is because of random error. It is evident from eq. (1) that any observed difference or error (e) will have a random component (δ) which limits our ability to estimate Δ. Just to (reliably) <u>detect</u> systematic error, it can be shown for normally-distributed random errors, that Δ must exceed SE by a factor of 4 or more (28). In order to detect a systematic error which is comparable to the standard deviation (σ), one therefore needs at least 15 observations.

Figure 2 indicates in a different way the difficulty in detecting sources of error. The solid curve shows the detection limit for bias (Δ) relative to the standard deviation (σ) as a function of the number of observations. The dotted curve gives the same type of information for another common problem: extraneous random error (σ_e) additional to the Poisson component in counting experiments (6). In this case, if the additional random error is twice the Poisson component one must have ten observations to demonstrate its existence. If the two are comparable, 47 observations suffice; and if the additional error is half the Poisson error, several hundred observations are required. Incidentally, the <u>same</u> (dotted) curve applies to the detection of the interlaboratory error (corresponding to σ_e) for a group of laboratories having comparable intralaboratory imprecision (corresponding to σ).

Clearly, in the absence of a very large number of measurements and long term stability, one cannot empirically (through intercomparison) establish error bounds (Δ or σ_e) much smaller than the standard deviation <u>of a single measurement</u> (σ). There is no substitute, however, for intercomparison and replication for the detection of unanticipated blunders or bias or lack of control which is relatively large compared to the standard deviation.

SOURCES OF SYSTEMATIC ERROR

The most effective way to identify and control sources of bias in chemical analysis is to treat the CMP as a carefully-defined system. Critical analysis of the individual steps and their linkage will then make it possible to estimate individual bias components as well as on overall propagated errors for systematic error. For this purpose a generalized flow diagram is given in figure 3. In the following paragraphs we shall examine each of the steps for possible systematic error contributions. A

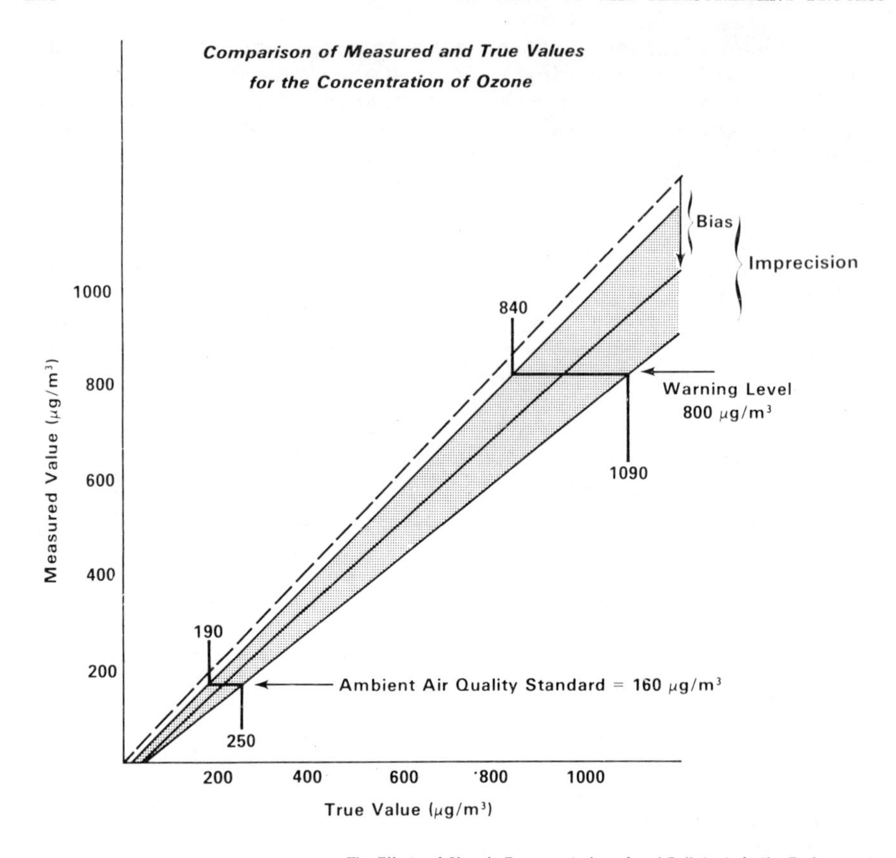

The Effects of Chronic Exposure to Low–Level Pollutants in the Environment

Figure 1. Results of a collaborative test of the EPA reference method for ambient ozone (2). Dashed line indicates the true value.

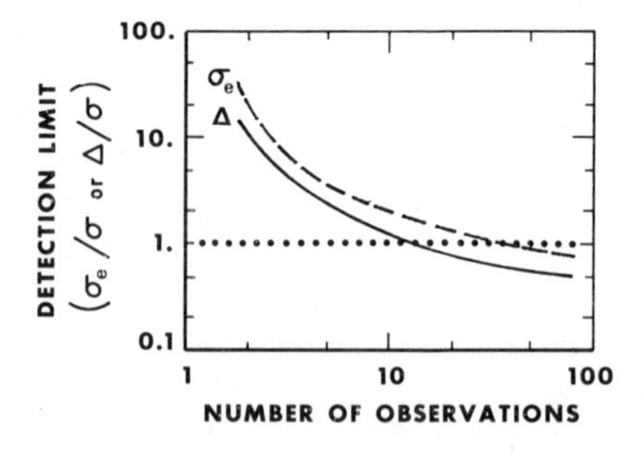

Figure 2. Detection limits vs number of observations for extraneous random error (σ_e, dashed curve) and systematic error (Δ, solid curve)

measurement process cannot be said to exist in the absence of control (7). When control is achieved both the central value and the variability are stable. Under such circumstances the error can be completely defined by a fixed systematic component (the bias) and a random component having constant standard deviation. Such bias may be the result of fixed mistakes or blunders in experiment or theory, or it may arise from converted random errors--i.e., errors which occur in a random fashion but which remain fixed because one or more steps of the CMP are not repeated.

If the systematic error is not constant, it becomes impossible to generate meaningful uncertainty bounds for experimental data. Lack of control may arise through carelessness or erratic blunders, such as transcription errors. These may be exposed via replication. Alternately, nonrandom variations may come about from the effects of systematic trends in uncontrolled variables (such as barometric pressure), or from unanticipated effects of seemingly remote factors. (Such effects are not necessarily a nuisance. They may provide an opportunity for discovery--as in the case of variations in the calibration curve for radiocarbon dating induced by the activities of man and various geophysical and climatic phenomena (8).)

The assessment of whether a measurement process is in control is frequently accomplished through the use of control charts--a technique which has been thoroughly discussed above. The control chart, of course, merely signals instability; it does not generally compensate for it. In order to achieve control, the experimenter must identify and either stabilize or correct for sources of erratic behavior. When variables cannot be held constant, it is often effective to correct for changes by means of an internal or external standard. Figure 4 gives one such example (39). Because of the extremely low concentrations present it was necessary to measure a sample of radioactive ^{37}Ar over a period of a month, during which time there was about a 10 percent drift in gain. Though it was not possible to prevent the drift, which came from slowly changing proportional counter gas composition, it was possible to correct for it. This was accomplished with an external monitor--an x-ray source which simulated the response of the detector to the sample radiation.

Sample Validity

Among the more serious problems affecting the sample are contamination, heterogeneity and instability. Contamination will be discussed below. The most likely consequence of heterogeneity is a nonrepresentative sample. Quite often one can observe a major difference in sample composition with amount taken for analysis. For example, the appearance of severe heterogeneity among trace elements in Orchard Leaves (SRM #1571)

	CRITICAL ASPECTS
X	Existence of CMP (definition, control)
Sampling	Homogeneity, contamination, stability
Separation (sample prep.)	Recovery, contamination
Measurement	Calibration, resolution
Evaluation	Model, error structure
$\hat{X}, e_{\pm}$	Adequate reporting Meaningful error bounds

Figure 3. The chemical measurement process—flow diagram

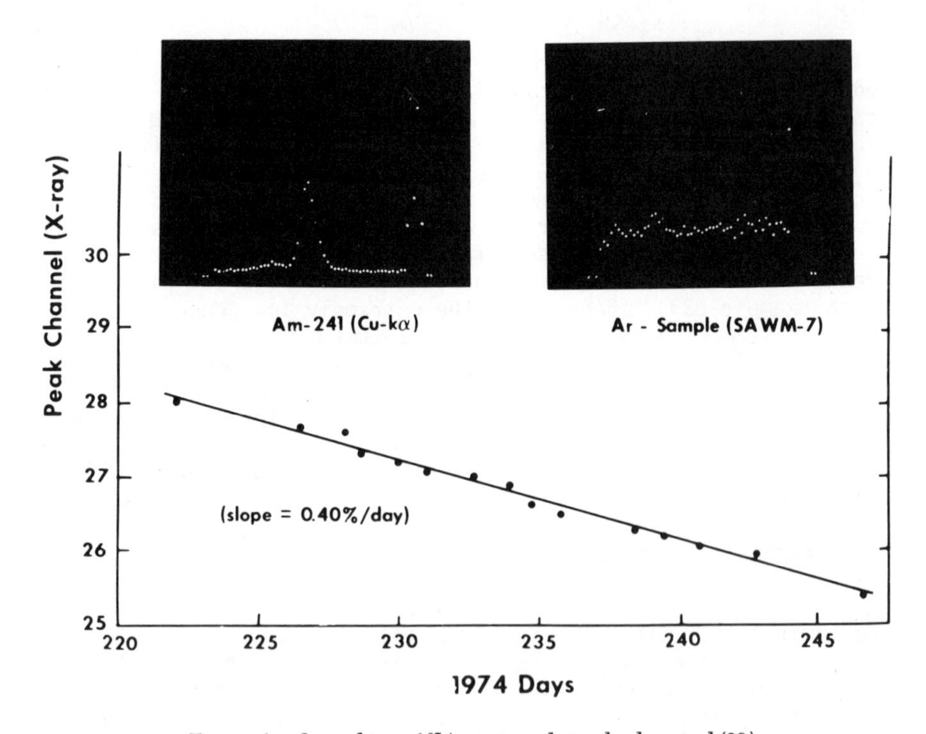

Figure 4. Quenching of ^{37}*Ar; external standard control* (39)

was demonstrated when 10 mg rather than the recommended 250 mg samples were taken (9). Such heterogeneity depends, of course, upon sample type. (The authors of ref. (10) noted the "extreme homogeneity" of trace elements in fresh liver.) Also, homogeneity of some elements cannot assure the same for others. For certain methods of analysis, sample homogeneity requirements are indeed stringent. Electron probe microanalysis, for example, requires standards of experimentally demonstrated microhomogeneity (10).

Another major source of systematic error relates to the change of sample composition with time. Aerosols, for example, are known to be susceptible to moisture and gaseous contaminants. Spurious sulfate results have been obtained from the gradual oxidation of SO_2 on air filters (11) and H_2SO_4 aerosol results have been falsely low because of partial neutralization by traces of NH_3 in laboratory air (12). Loss of trace species from aqueous samples at container walls (adsorption or diffusion) is another common source of instability. This is particularly marked for heavy, volatile elements such as mercury (13).

The Blank

Figure 5 illustrates a systematic error that is troublesome in the first two steps of the CMP: that is the occurrence of an unmeasured blank. A significant difference is shown between a simple solution of lead and the apparent lead content in whole blood (14). When the deviations at these low levels of concentration are all positive, a good supposition is a blank problem from contamination of reagents used to prepare the sample of whole blood but not used for the aqueous solution.

A common pitfall in trace analysis is insufficient attention to the variability of the blank. If variability due to contamination is such that it may play an important part in the setting of uncertainty bounds or detection limits, some caution is necessary in interpreting the result of just one or a few blank observations (19). Results such as those quoted above show the danger of blindly assuming that the relative range of the blank is no more than 10 percent, 100 percent, a factor of 2 or even a factor of 10^6 (15). Thus, even if the blank is ten times smaller than the signal of interest, its variability _must_ be measured. If this is accomplished, for example, by examining the difference of just two experimental blanks, there is a significant chance that the actual range of the blanks will exceed that measured difference by a factor of 25, under the best of circumstances (normally-distributed blanks; 95 percent tolerance interval)! Giving up the assumption of normality, but requiring the blank to be under (statistical) control, one can be fairly (95 percent) certain that half of the blanks will fall within the range of 8 observations, or 90 percent of them within the range of _47_ observations!

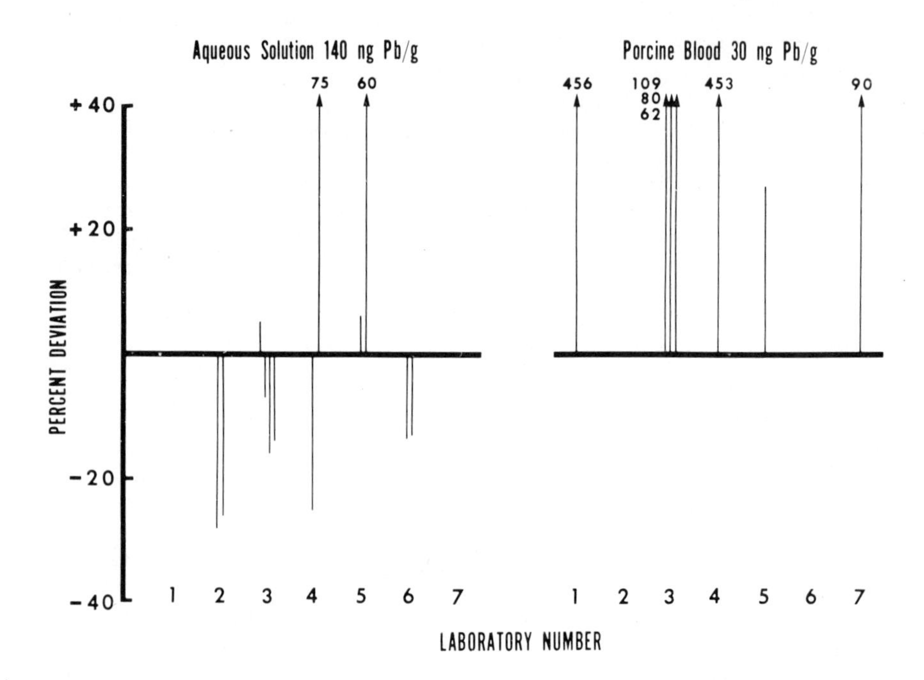

National Bureau of Standards Special Publication

Figure 5. Comparison of interlaboratory Pb results for an aqueous standard vs. whole blood (14)

Sample Preparation

Besides problems with the blank, great care must be taken when performing physical or chemical separations of components in a sample. If the recovery of the desired component is not quantitative,--*e.g.*, loss of volatile components during sample dissolution--serious systematic error may result (16).

The recovery factor presents the same opportunity to err as does the instrument calibration factor--namely, the assumption that the (average) yield is quantitative or constant and that its variability (relative standard deviation) is fixed. Such fixed values may be deduced by assumption, a theoretical model (solubility product, partition coefficient, ...) or better still, by a few measurements on pure solutions. The trap is laid: as soon as varying concentrations and complex samples are encountered low and fluctuating yields will occur.

One of the most reliable means for eliminating bias due to nonquantitative separation is isotope dilution. Provided that the diluting isotope is added at the earliest possible stage and that complete isotopic mixing takes place, this technique is capable of very high accuracy. The art has perhaps reached its ultimate level at the hands of skilled chemical mass spectrometrists, who have succeeded in measuring isotope ratios with uncertainties of only 0.03 percent (17,18).

Measurement

The measurement step provides many chances for error. Operator bias, for example, commonly occurs in the making or recording of observations, as shown in figures 6 and 7 (19,20). Results of 1,000 weighings (fig. 6) show that operators favor the values of 0 and 5 for the last digit, and that even numbers tend to be favored over odd numbers. From 1,510 buret readings (fig. 7), on the other hand, one can observe that small numbers are favored over large numbers. The possibility of operator bias is, perhaps, sufficient justification for considering computer control for such types of processes. (Since even computers are programmed and run by the operator of the instruments, however, the threat of errors (blunders) of this type is only reduced, not eliminated.)

Two of the most important characteristics of analytical measurements are the calibration function and instrumental resolution. To assume that the calibration factor is constant, independent of the nature (matrix) or concentration of the sample, is to invite bias. It is in the calibration factor, together with recovery factors, that "real" samples differ most strikingly from pure solutions. Aside from the use of sound theoretical or semi-empirical correction formulas, the most reliable method to assure a correct calibration is the use of an

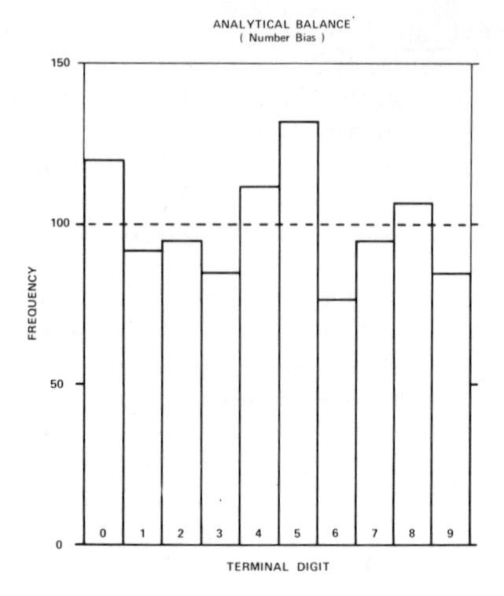

Figure 6. Operator bias—analytical balance. Histogram depicts observed terminal (estimated) digit distribution for 1000 student weighings. Dashed line indicates expected distribution. (Data from Ref. 20).

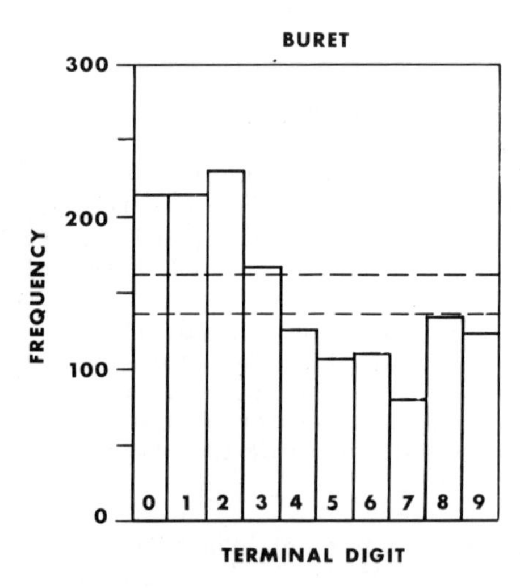

Figure 7. Operator bias—buret reading. Histogram depicts terminal digit distribution for 1510 student observations. Dashed lines delimit the 95% confidence interval for a uniform distribution. (Data from Ref. 20).

internal standard. By adding to the sample a known aliquot of the substance being measured, one can translate the differential response into an effective calibration factor for the actual sample at hand (21).

Instrumental resolution, just like chemical or physical "resolution"--*i.e.*, separation--is one of the most important means of preventing systematic error from unanticipated components or ill-defined spectral features. Some of the penalties from inadequate resolution will be examined below in our discussion of data evaluation. When dealing with complex materials containing potentially interfering species, however, a small investment in increased *chemical* resolution will be well repaid in decreased bias.

Data Evaluation

Although the last two steps of the CMP do not necessarily involve any work in the chemical laboratory, they are nevertheless an integral part of the overall measurement system and thus must be recognized as potential contributors of systematic error. In fact, perfectly valid sampling, chemistry and instrumental measurement can be rendered meaningless by faulty evaluation or reporting. This potential for inaccurate data evaluation has been recognized recently in an international comparison devoted strictly to the evaluation (and reporting) phase of gamma-ray spectroscopy (22).

Errors which are due to the differences between pure solutions and complex samples often remain latent until the evaluation stage. Systematic error can be minimized provided that such differences--connected with the blank, matrix effects, component interference--are adequately recognized in the evaluation process. (The possibility of making proper corrections may, of course, depend upon the prior introduction of a recovery tracer or use of a high resolution measuring device.)

For single component measurements a common source of evaluation bias is the assumed calibration "constant." Matrix corrections represent one area where the analyst must correctly adjust this factor (10,23). The other relates to the functional relationship assumed between the quantitative response of an analytical chemistry measurement system and the composition of standards. Many times the relationship is linear or at least it appears to be so. However, one soon learns that he can define a fit to a mathematical model in a variety of ways. It is in this process of determining whether the model adequately represents the experimental data, that systematic errors can arise. A common but potentially misleading calibration procedure is fitting a straight line to the data and the subsequent examination of a test statistic to assess the goodness of fit.

An example is the calibration of linearity of the energy scale of a Ge(Li) γ-ray detector. Figure 8 shows a linear fit where the residual relative standard deviation (a measure of fit) was less than 0.1 percent, and the correlation coefficient (a measure of linearity) was 0.9999.

However, through more detailed examination we found that the fit was not really adequate when compared with that expected on the basis of Poisson counting statistics. A very informative way to evaluate the fit is to observe the plot of residuals (algebraic difference between the experimental data points and the fitted mathematical model vs γ-ray energy). One can see in the figure that the X's are not distributed randomly about zero. In fact, errors in both the calibration function and the tabulated standard energies were detected; the corrected results are represented by the dots. Although great improvement was obtained, one can see by inspection that there is a slight decrease in the spread of the residuals at higher channel numbers. This indicates a possible additional problem that might warrant further study.

Multicomponent methods of analysis often suffer bias from inadequate resolution. The problem of accurately resolving obviously overlapping peaks, such as those shown in figure 9, has received considerable attention in the spectroscopic and chromatographic literature (24). Not so well appreciated, however, is the fact that significant systematic error may be introduced when an interfering peak is present but not apparent, and hence excluded from the data reduction model (25). The magnitude of the resulting bias, when an undetected peak lies buried within the peak of interest is shown in figure 10. It is a surprising result that the level of error can be so large and still go undetected. Plotted in the figure is the ratio of the systematic error to the standard deviation of the estimated area of a (Guassian) peak as a function of its separation from a neighboring (undetected) peak. It can be seen that if the overlap is equal to or less than the half width, very large systematic error can result (26).

Improved instrumental resolution may eliminate the above pitfall. In fact, advanced instrumentation may reveal quite a surprising degree of complexity; figure 11 shows the structure actually contained in the apparent γ-ray doublet of figure 9.

Reporting Results and Uncertainties

Among the results reported in a recent trace analysis laboratory intercomparison of an NBS Standard Reference Material (SRM 1577, bovine liver), one finds the following:

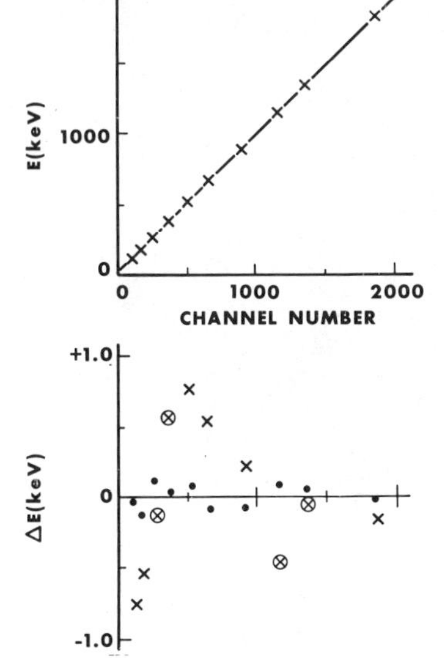

Figure 8. a. Gamma–ray calibration curve: energy (keV) vs. channel number; b. residuals (ΔkeV) from gamma–ray calibration curve vs. channel number. (x = linear function; • = cubic function; ○ = bad physical input data [tabulated γ–energy].)

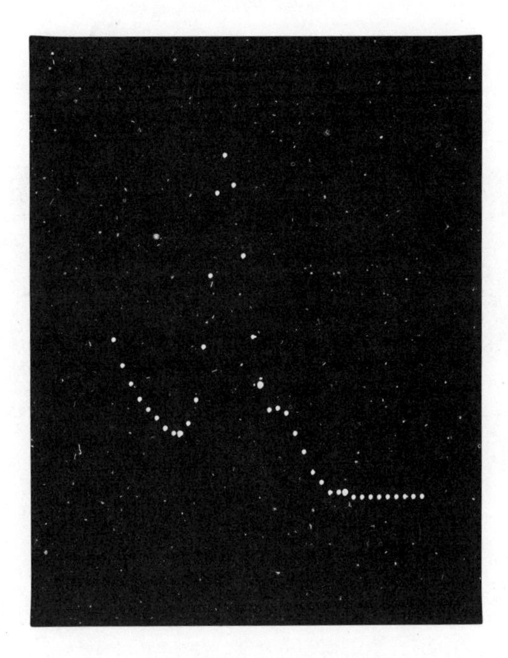

Figure 9. Gamma–ray spectrum from Bremsstrahlung-activated gold: NaI(Tl)detector

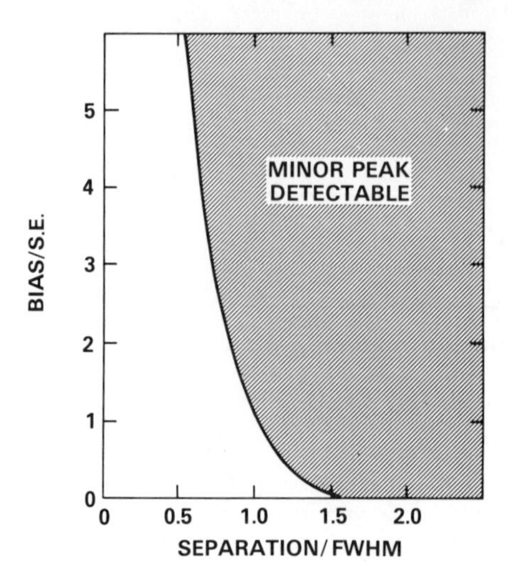

Figure 10. Model error bias. Curve shows the (maximum) bias in the estimated area of the major peak in a spectral doublet when an undetected minor peak is omitted from the mathematical model. The minor peak is not detectable if it lies below the solid curve.

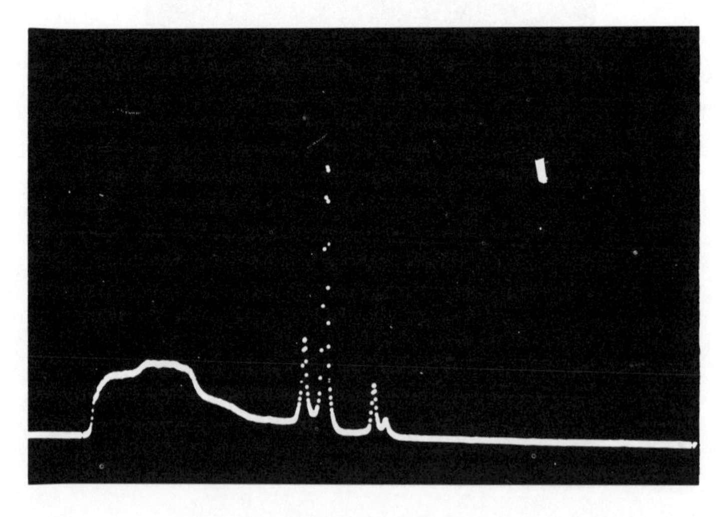

Figure 11. Gamma-ray spectrum from Bremsstrahlung - activated gold: Ge(Li)detector

Hg-content (μg/g): 0.09
<0.2
0.006 ± 0.0006

In the absence of additional information, interpretation of the results individually or collectively is obviously impossible. One has no idea of the uncertainty associated with the first result, and the meaning of the upper limit for the second is unknown. Besides the mismatch of decimal digits, the result reported by the third laboratory includes an uncertainty which allows several interpretations (27).

Clearly, results of good experiments inadequately communicated are worthless. A full, verbal explanation of all reported quantities and uncertainties is always in order. We recommend, as a minimum set, the following information:

(a) The result which was actually computed--even if it be negative.

(b) For "nonsignificant" or nondetected results:

(i) the detection criterion

(ii) an upper limit, together with its meaning.

(c) The estimated standard error.

(i) number of observations (n) and the number of degrees of freedom (ν), if estimated by replication. (ν = n-1 for simple, one-parameter estimates or averages.)

(ii) source of the estimate, if no replication--*e.g.*, prior experience Poisson statistics, ...

(d) The estimated bounds for systematic error (not necessarily symmetric), and method of estimation--*e.g.*, converted random error (nonrepeated step, as calibration), comparison with reference material or alternative method, propagation of bias bounds of intrinsic CMP steps, *etc*.

Finally, the overall uncertainty of the experimental result may be given as a combination of the random and systematic components, provided the individual components ((c) and (d), above) and the explicit combination recipe is included. Because of lack of knowledge concerning error distributions and because of the somewhat subjective nature of inferred systematic error bounds (4), the conservative approach is preferred: simple summation of the random and systematic error bounds, where the former is equal to the standard error multiplied by a given percentage of Student's-t distribution. References (28-30) contain critical information on the estimation and treatment of uncertainty bounds.

DIAGNOSTIC TECHNIQUES

The task of the analyst is incomplete until systematic error or blunders, which have been detected, have been thoroughly examined and eliminated.

Important progress in the Analysis of Blunders (ANOB) may be made through the use of certain numerical and graphical procedures which are just now being developed (31,32,33). Such techniques have been discussed in detail above; the following examples illustrate their application to specific analytical problems involving systematic error. The numerical methods utilize "resistive" and "robust" statistics which are generally immune to the effects of outliers and assumptions concerning random error distributions, respectively. The graphical methods include the use of histograms, residual plots and correlation diagrams. Such one- or two-dimensional displays of the data can be an exceptionally efficient way to pinpoint erratic blunders and sources of bias. The purpose of such procedures is error diagnosis leading to improved experiments--not the repair of faulty data.

When assumptions--such as normality, linear calibration curves, negligible outlier or bias occurrence--are questionable, then resistive, robust and distribution-free methods are in order. Among the most convenient for the analyst, who often has a relatively small number of observations, is the median and its confidence interval (see Table I and ref. (38)). Both quantities can be determined from a set of data simply by ordering, without computation. Furthermore, the estimate for the confidence interval is resistant to the effects of an outlier once the number of observations exceeds 8. (For $n = 6,7,8$ the 95 percent confidence interval for the median is just equal to the range.) With respect to the assessment of variability, a convenient method for estimating the standard deviation from a set of residuals is to take the median (unsigned) residual. Considerable protection against bad data is then afforded, especially as compared to the use of the (conventional) root mean squared residual.

1-Dimensional Plots - Residuals

Systematic errors were revealed above by means of histograms (figs. 6,7) and residual plots (fig. 8). Such plots are always helpful, but they are certainly called for whenever χ^2 (computed to test regression model or distributional accuracy) is unacceptable or when the estimated variance (s^2) from model fitting is inconsistent with the expected variance (σ^2), as in fig. 8.

Table I. Confidence interval for the median[a,b,c] (distribution-free)

n	k	n	k
6-8	1	20-22	6
9-11	2	23-24	7
12-14	3	25-27	8
15-16	4	28-29	9
17-19	5	30-32	10

[a] Entries derived from table A-25 of reference (38).

[b] If observations are ordered:

$x_1, x_2, \ldots x_n$, the confidence interval

($\alpha \leq .05$) equals x_k to x_{n-k+1}

[c] CI = w (range) for n = 6,7,8

CI ≈ Interquartile range (mid-50 percent of the observations) for n = 12 to 24.

Such inconsistency ($s^2 >> \sigma^2$) led to the application of resistive and residual pattern techniques for the diagnosis of systematic error in the XRF analysis of CaO (34). A straight line fit to the calibration curve gave an estimated RSD of 1.2 percent, seven times the value expected from Poisson counting statistics (see table II). Figure 12a displays the normalized residuals resulting from the unweighted least-squares fit of a linear calibration curve to the data. Two peculiarities are suggested by this plot: (a) A slightly nonrandom pattern is apparent. (b) The largest residuals are too small--there is scarcely an excursion beyond ±1s.

An alternative method of computation, using the same model but incorporating the observed background ($\hat{b}$) and Poisson statistics, was then performed. To provide some measure of insensitivity to outliers residuals were then calculated from the median value of the estimated calibration constant ($\hat{A}_m$). The computation formulas and numerical results are shown in table II; the resulting residual plot is given in figure 12b. The pattern remains, but now we see significant excursions beyond the (Poisson) control limits, with the first measurement far worse than the remainder. We tentatively mark the first result, therefore, as a blunder. Note that this initial error was revealed only through the use of the resistive technique; it was completely masked when conventional least squares was applied.

Table II.[a] Calcium--x-ray fluorescence calibration curve model: $y = b + Ax + e$

x(% CaO)	y(counts/500s)	$\hat{\hat{A}}$
0.202	83.5k	387.6k
0.275	123.2k	428.5k
0.719	307.7k	420.5k
0.812	342.4k	415.1k (median)
1.636	682.1k	413.8k
2.047	854.3k	414.8k
3.969	1661.2k	417.2k

Least Squares	Poisson
$\hat{b} = 3620 \pm 2360$	$\hat{\hat{b}} = 5240 \pm 70$
$\hat{A} = 417.1 \pm 1.1k$	$\hat{\hat{A}}_m = 415.1 \pm 0.7k$
$s = 1.2\%$	$\bar{\sigma} = 0.17\%$

$s/\bar{\sigma} \simeq 7.$

[a] $\hat{b}$, $\hat{A}$, result from fitting the model. $\hat{\hat{b}}$ = observed background; $\hat{\hat{A}}_i = (y_i - \hat{\hat{b}})/x_i$.

The pattern among the remaining residuals in figure 12b does not exhibit the smoothness expected from a systematic effect. This was clarified by further inquiry, however, when we learned that the samples were loaded on a rotary sample wheel in order of increasing calcium concentration. Taking this into account, and deleting the presumed outlier, we replotted the residuals as a function of sample position in figure 12c. The relatively symmetric, smooth pattern is quite consistent with sample wheel distortion which produces varying response through sample-detector distance variations. Construction of an improved sample wheel eliminated this source of systematic error, yielding an experimental imprecision of just 0.2 percent (35).

Another important point should be made about this measurement method. Had it not been possible to adequately reduce the wobble from the turntable, it would have been best to randomize the positions of the CaO samples. This would transform a potential systematic error into a random error and of course reduce the precision of the measurement. Additional testing of the magnitude of the effect of the wobble would then be possible by plotting (or correlating) the residuals with position (θ). Alternatively, a plot of residuals *vs* CaO concentration might reveal concentration-dependent systematic error.

2-Dimensional Plots - Correlation

Interlaboratory comparisons provide an extremely powerful method for revealing systematic error. Using a technique originally conceived by Youden (36), one can effectively distinguish random error, laboratory bias and erratic blunders by a simple two-dimensional correlation technique. As originally devised, the method required the analysis of two similar samples by each of the participating laboratories. As a preliminary diagnostic method, however, it is useful even with the relaxation of the similarity constraint.

To illustrate the potential of the method, figure 13 shows vanadium results from 10 laboratories who participated in an NBS-EPA trace element intercomparison (37). Each laboratory was provided with samples of two NBS-SRM's (SRM #1632, trace elements in coal and SRM #1633, trace elements in coal fly ash). The test statistic which led us to explore graphic systematic error diagnosis was once again the variance ratio: the interlaboratory error had been computed to be several times the intralaboratory component.

The results are striking. Despite the appearance of a "fitted line" in figure 13, the points shown and the line are totally independent! The line has no free parameters. Its slope (on the log-log plot) was fixed at +45° based on the hypothesis of a multiplicative bias model--*i.e.*, biased calibration factors. The location of the line was fixed by the known vanadium concentrations (dashed rectangle) for the two SRM's. (The concentrations were known to NBS, *not* to the participants.) Nine of the results shown (solid circles) involved replication; the tenth (cross) did not.

The coherence of the replicated results to the theoretical line confirm the assumption of multiplicative bias, and their deviations from the line are consistent with the intralaboratory imprecision. It follows that use of the SRM's to expose the individual calibration biases has the potential for bringing all laboratories within the bounds expected from the intralaboratory errors--an improvement in reproducibility by at least a factor of two. The non-repeated measurement (cross) demonstrates also the vulnerability of isolated observations to erratic blunders. Displacement from the line in this case indicates a mistake in the analysis of the fly ash sample or the coal sample, or both.

SUMMARY

This chapter has been directed toward four key aspects of systematic error in chemical analysis: (a) the serious conse-

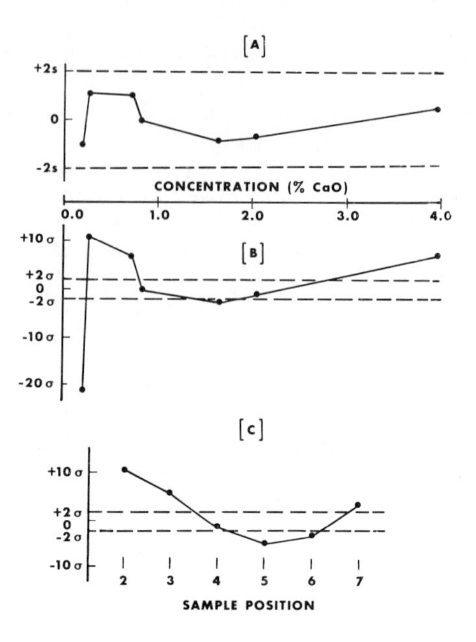

Figure 12. Residual diagnosis of systematic error in CaO XRF analysis. (a) Residual pattern (y–y) following least squares fitting of a linear calibration curve, (b) normalized Poisson residuals (A–A) using resistant method (A_m), (c) normalized Poisson residuals ($A–A_m$) recalculated following deletion of the first point using sample position as the independent variable.

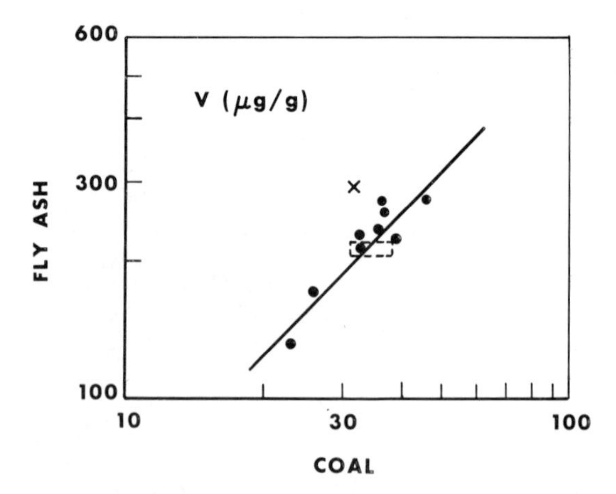

Figure 13. Results of NBS–EPA vanadium intercomparison. X represents the result which lacks replication.

quences of inaccuracy in the external use of analytical results, (b) the essential requirement of systematic error detection or accuracy verification via laboratory or method intercomparison, (c) the systems analytic approach to the CMP as the only reliable, organized way to anticipate the origin, magnitude and flow of systematic error, (d) the power of numerical and graphical diagnostic techniques, which are rapid and relatively immune to assumptions and blunders, for exposing the particular nature of systematic errors following their detection.

It is clear that the need for accuracy in analytical chemistry continues to increase along with our understanding of the importance of the composition of materials. Errors in such measurements can lead to incorrect decisions in fields ranging from environmental protection to the diagnosis of disease.

Accuracy assurance can benefit from the use of reference materials of known composition, so long as careful iterative feedback of information among laboratories is used to eliminate methodic errors. In addition it is vital to investigate, on an individual-analyst basis, possible sources of errors in each of the steps of the specific CMP, for it is evident (fig. 2) that bias which is comparable to or smaller than the imprecision can easily escape empirical detection.

Often we understand where many of the errors arise. Lack of control, sampling, dissolution, chemical separation and purification, and instrumental errors come immediately to mind.

Table III. Assumption Limitations

(negligible bias + estimate imprecision → meaningful results)

Random Error:	Poisson deviations from normality ($N < 30$) Random component of systematic error sources Random errors in corrections for systematic errors
Systematic Error:	Sampling and sample preparation (recovery) Blank, interference, and contamination Improper calibration and/or standards Matrix effect--particle size and composition, enhancement, adsorption, and scattering Inaccurate data reduction models or correction formulas (assumed parameters, functional relations) Blunders and faulty reporting

However, reliable data evaluation and reporting are no less important. Great care must be taken to employ mathematical procedures for data reduction which reflect the actual physicochemical processes of the entire analytical system. Inadequate reporting, at best, can lead to misinterpretation of results and consequent erroneous decisions. In order to give a capsule summary of all of these factors, we show in table III (27) the deviations from assumptions which are most likely to lead to unanticipated error.

Terms and Symbols Used in Text and Figures

CMP = Chemical Measurement Process

n = number of observations

ν = degrees of freedom = number observations minus number of estimated parameters (unknowns)

τ = true value (if known)

$\bar{x}$ = experimental result (mean)

e = error = result - truth = $\bar{x} - \tau$

$e_{\pm}$ = uncertainty bounds in e

δ = random error in $\bar{x}$

δ_M = maximum likely (absolute value) random error--typically taken as two to three times the standard error

σ = standard deviation (single observation)

z = random normal deviate

SE = Standard Error, standard deviation of mean = $\sigma/\sqrt{n}$

σ_e = extraneous random error

Δ = systematic error in $\bar{x}$--*i.e.*, nonrandom component of error equals constant bias when the CMP is in control

$\hat{\Delta}_{\pm}$ = estimated bounds for Δ (result of experimental comparison)

$\tilde{\Delta}_{\pm}$ = inferred bounds for Δ (result of CMP-structure and scientific judgment)

Literature Cited

1. New York Times, Mar. 28, 1976, p E-9.
2. McNesby, J., Testimony on Assessment - Need for Standards, at Hearing on The Effects of Chronic Exposure to Low- Level Pollutants in the Environment before the Subcommittee on the Environment and the Atmosphere, U.S. Congress (1975).
3. "Some Europium-152 Gamma-Ray Probabilities," Radioactivity Section, NBS Oct. 1976.
4. Taylor, B. N., Parker, W. H. and Langenberg, D. N., The Fundamental Constants and Quantum Electrodynamics, Academic Press, New York (1969). [See especially p 20].
5. Yolken, H. T., The Role of Standard Reference Materials in Environmental Monitoring, NBS Spec. Publ. 409, Marine Pollution Monitoring (Petroleum), Proceedings of a Symposium held at NBS, Gaithersburg, MD, May 13-17, 1974.
6. Currie, L. A., The Limit of Precision in Nuclear and Analytical Chemistry, Nucl. Instr. Meth., (1972), 100, 397.
7. Eisenhart, C., Realistic Evaluation of the Precision and Accuracy of Instrument Calibration Systems, Journal of Research NBS, (1963), Vol. 67C, No. 2, p 161.
8. Olsson, I. U., Editor, Radiocarbon Variations and Absolute Chronology, Proceedings of the 12th Nobel Symposium held at the Institute of Physics at Uppsala University, Wiley Interscience (1970).
9. Campbell, J. L., Orr, B. H., Herman, A. W., McNelles, L. A., Thomson, J. A. and Cook, W. B., Trace Element Analysis of Fluids by Proton-Induced X-Ray Fluorescence Spectrometry, Anal. Chem., (1975), 47, 1542.
10. Heinrich, K. F. J., Common Source of Error in Electron Probe Microanalysis, Advances in X-Ray Analysis, (1968), Vol. 11, p 40.
11. Lee, R. E., A Sampling Anomaly in the Determination of Atmospheric Sulfate Concentration, Am. Ind. Hyg. Assoc. J., (1966), 27, 266.
12. Charlson, R. J., Vanderpol, A. H., Covert, D. S., Waggoner, A. P. and Ahlquist, N. C., $H_2SO_4/(NH_4)_2SO_4$ Background Aerosol: Optical Detection in St. Louis Region, Atmospheric Environment, (1974), Vol. 8, p 1257.
13. Burrows, W. D. and Krenkel, P. A., Loss of Methylmercury-203 from Water, Anal. Chem., (1974), 46, 1613.
14. Murphy, T. J., The Role of Analytical Blank in Accurate Trace Analysis, NBS Spec. Publ. 422, (1976), Vol. II, p 509.
15. Robertson, D. E., Role of Contamination in Trace Element Analysis of Sea Water, Anal. Chem., (1968), 40, 1067.
16. Apt, K. E. and Gladney, E. S., Loss of Osmium During Fusion of Geological Materials, Anal. Chem., (1975), Vol. 47, 1484.

17. Barnes, I. L., Murphy, T. J., Gramlich, J. W. and Shields, W. R., Lead Separation by Anodic Deposition and Isotope Ratio Mass Spectrometry of Microgram and Smaller Samples of Lead, Anal. Chem., (1973), 45, 1881.
18. Moore, L. J., Machlan, L. A., Shields, W. R. and Garner, E. L., Internal Normalization Techniques for High Accuracy Isotope Dilution Analyses - Application to Molybdenum and Nickel in Standard Reference Materials, Anal. Chem., (1974), 46, 1082.
19. Hume, D., Pitfalls in the Determination of Environmental Trace Metals, Progress in Anal. Chem., Vol. 5, "Chemical Analysis of the Environment," Plenum Press (1973).
20. Laitinen, H. A. and Harris, W. E., Chemical Analysis, McGraw-Hill Book Company, Second Edition (1975).
21. Intersociety Committee, Tentative Method of Analysis for Vanadium Content of Atmospheric Particulate Matter by Atomic Absorption Spectroscopy, Health Lab. Sci., (1974), 11, No. 3, 240.
22. International Atomic Energy Agency, "Intercomparison of Methods for Processing Ge(Li) Gamma-Ray Spectra," H. Houtermans and R. M. Parr, Analytical Quality Control Services (1976).
23. Nargolwalla, S. S. and Przybylowicz, E. P., Activation Analysis With Neutron Generators, Sources and Reduction of Systematic Error, Chap. 6, p 255, John Wiley and Sons (1973).
24. Blackburn, J. A., Editor, Spectral Analysis: Methods and Techniques, Marcel Dekker, New York, NY (1970).
25. Currie, L. A., The Discovery of Errors in the Detection of Trace Components in Gamma Spectral Analysis, Modern Trends in Activation Analysis, Vol. II, DeVoe, J. R. and LaFleur, P. D., Editors, p 1215, NBS Spec. Publ. 312 (1968).
26. Currie, L. A., Sources of Error and the Approach to Accuracy in Analytical Chemistry, Chap. 4 in Vol. I, Treatise on Analytical Chemistry, P. Elving and I. M. Kolthoff, Editors, J. Wiley and Sons, New York, NY (1977).
27. Currie, L. A., Detection and Quantitation in X-Ray Fluorescence Spectrometry, Chap. 23 in X-Ray Fluorescence Methods for Analysis of Environmental Samples, T. Dzubay, Editor, Ann Arbor Science Pub., Ann Arbor (1976).
28. Ku, H. H., Precision Measurement and Calibration - Statistical Concepts and Procedures, NBS Spec. Publ. 300, Vol. I, Superintendent of Documents, U.S. Government Printing Office, Washington, DC (1969).
29. Eisenhart, C., Expression of the Uncertainties of Final Results, Science, (1968), 160, 1201.
30. "Round-Table Discussion on Statement of Data and Errors," Nuclear Instr. Meth., (1973), 112, 391.

31. Lide, D. R., Jr. and Paul, M. A., Critical Evaluation of Chemical and Physical Structural Information, Chap. I, "Analysis of Experimental Data," Conference Proceedings Dartmouth College, National Academy of Sciences (1973)
32. Tukey, J. W., Exploratory Data Analysis, Addison-Wesley (1977).
33. Draper, N. H. and Smith, H., Applied Regression Analysis, Chap. 3, "The Analysis of Residuals," John Wiley and Sons (1966).
34. Experimental Data; Courtesy of P. Pella, NBS (1976).
35. Breiter, D., Personal Communication (1977).
36. Youden, W. J., The Sample, The Procedure, and the Laboratory, Anal. Chem., (1960), 32 [13], 23A.
37. EPA-NBS Interlaboratory Comparison for Chemical Elements in Coal, Fly Ash, Fuel Oil and Gasoline (1973).
38. Dixon, W. J. and Massey, F. J., Jr., Introduction to Statistical Analysis, McGraw-Hill Book Company, Inc., Third Edition (1969).
39. "Natural Production Rate and Atmospheric Concentration of Ar-37", L. A. Currie, R. M. Lindstrom, J. F. Barkley and P. S. Shoenfeld, Technical Note of the National Bureau of Standards to be published; 1977.

4

Role of Reference Materials and Reference Methods in the Measurement Process

GEORGE A. URIANO and J. PAUL CALI

Office of Standard Reference Materials, Institute for Materials Research, National Bureau of Standards, Washington, DC 20234

This paper is concerned with the role of reference materials and reference methods in the measurement process. Reference materials and reference methods are considered to be two vital components of measurement systems needed to assure the accuracy and compatibility of measurements.

The views expressed in this paper are an outgrowth of ideas and concepts expressed recently in a number of publications (1,2,3) by J. Paul Cali and other members of the NBS staff. In particular, NBS Monograph 148 provides extensive background information concerning the use of reference methods and reference materials. This monograph was written in response to a request from the Agency for International Development to provide assistance for improving the measurement systems of developing countries.

This paper consists of two parts. The first part is concerned with the achievement of measurement compatibility. Some general considerations concerning the use of reference materials and reference methods in the measurement process are discussed first. Reference materials and reference methods are seen to be two necessary but not always sufficient mechanisms for achieving measurement compatibility between laboratories on a national scale. The general discussion of measurement compatibility is aimed at providing the conceptual framework needed to examine two specific examples of how reference methods and/or reference materials have improved measurement systems for determining calcium in serum and nitrogen dioxide in ambient air. Evidence is also presented showing that the lack of adequate reference materials and/or reference methods is impeding the development of compatible, national measurement systems for the determination of trace amounts of mercury in water and for the measurement of trace chromium in biological matrices.

THE ACHIEVEMENT OF MEASUREMENT COMPATIBILITY

The Importance of Measurement Compatibility

Why do we make measurements? Measurements are important for a number of reasons. Measurements provide the basis for equity in trade and for settling disputes between buyer and seller or producer and user. Measurements are important to the industrial quality control process (4), and are used to assess and improve the reliability or performance of materials and systems. In recent years a number of social and/or political considerations have highlighted the importance of good measurements, particularly in the areas of medical diagnosis (5) as well as the enforcement of environmental (6) and occupational safety regulations. Finally, measurements provide us with the quantitative, scientific, and engineering data that serve as the language of science, allowing improved communication of information among the scientists of the world, even across language barriers.

What do we mean by the measurement process? The measurement process consists of at least two components. First, some type of a scale is needed that allows one to quantitatively estimate the value of an intrinsic or extrinsic property of a material or system. Second, a method for applying the scale to whatever property is being measured is needed. The end result of applying a method plus a scale is to arrive at a number that allows a definite value to be assigned to the property under consideration by means of a measurement-property relationship.

In all communications or transactions involving two or more parties, whether they be economic, socio-political or scientific, one of the critical steps in the transaction is that the parties involved agree on the results of the measurement and the meaning of the numbers associated with the measurement. This agreement should take into consideration any imprecision and inaccuracies inherent in the measurement process under consideration. If the measurements are in agreement, we say that they are compatible. Thus all people involved in communicating via the measurement can agree that the measurement is useful for whatever end purposes the measurement was made.

By definition, a measurement is accurate when the resulting numbers are both precise and free of any systematic error. Under these conditions compatibility between parties is not only possible but highly probable. This leads to the somewhat obvious but not always appreciated conclusion that measurement accuracy leads to measurement compatibility.

Mechanisms for Achieving Measurement Compatibility

There are a number of different mechanisms by which measurement compatibility may be achieved and accuracy transferred between two or more laboratories. For example, all parties might agree to use a reliable, stable, generated radio signal (e.g., the time signals transmitted by NBS radio station WWV) as the common reference point for assuring the compatibility of time measurements.

The use of reference data provides another means for assuring measurement compatibility. Temperature measurements can be made compatible through the use of reference data such as the melting point of pure substances. Compatibility of electrical conductivity measurements can also be achieved in a similar manner. The purity of the material is very crucial in such an application. Unless all parties are using (the same) "pure" materials, compatibility will not be assured.

A third way to achieve compatibility is through the use of reference materials as a transfer medium. In the broadest sense, a reference material is a material, device, or system, which has been constructed or modified in such a way that definitive numerical values can be associated with specific properties. The "material" might be an ozone generator that emits known amounts of ozone or a homogeneous metal alloy that contains a known amount of chromium.

By using reference materials, measurement compatibility can be achieved on the basis of precision alone, if all parties agree to use the same measurement methods and reference material. This mode of achieving compatibility is illustrated schematically in figure 1. Whether the laboratories are obtaining the "true value" or not is unimportant as long as the measurement is confined to the group of laboratories (A to D), all of which are using exactly the same methods and reference materials. Buyer-seller transactions for example can occur in a compatible manner provided we remain within the immediate universe of users i.e., among laboratories (A to D). However, if Laboratory A has to communicate with a laboratory outside this immediate domain (e.g., laboratory E) and that laboratory is using a different measurement method and reference material, then measurement compatibility may be difficult if not impossible to achieve if precision alone is the basis. This problem can usually be alleviated by achieving compatibility through accuracy rather than precision. In that case the recommended measurement methods and reference materials would have known uncertainties associated with them, i.e., they would be characterized in terms of reliable known values denoting both imprecision and systematic errors. One important mechanism for achieving compatibility on the basis of accuracy is through the use of reference materials and reference methods. This is the mechanism used in the determination of calcium in serum to be described later.

Accurate Measurement and True Values

This leads us to the concept of accurate measurement and true values. Let us perform the following thought experiment (illustrated in figure 2): We are to measure a specific property of a liquid substance, which is stable with respect to time and homogeneous with respect to spatial variations of the property being measured. Without getting into the philosophy of what one means by "true value"--let us all agree that there can be only one unique true value of this property--for example the copper content of the liquid.

Since measurement compatibility implies a series of users, suppose we have *N* labs each making measurements in a way that they all get the true

- All laboratories (A-D), in universe using same methods and reference materials

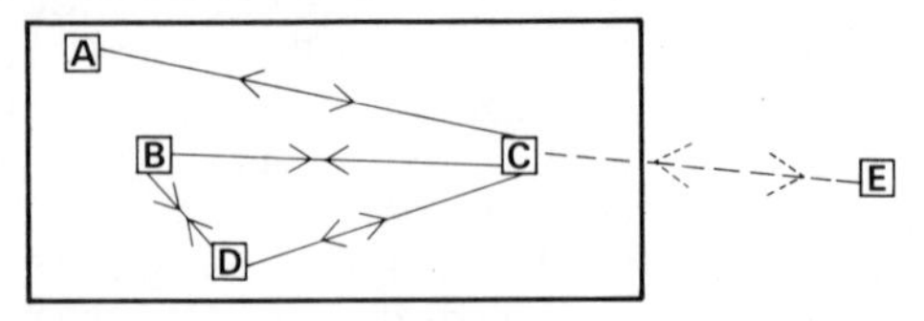

- Hence, compatibility within immediate universe of users
- What happens when "C" attempts to communicate with "E" who is using different methods and reference materials?

Figure 1. Schematic of how measurement compatibility may be achieved through precision alone, if all interacting laboratories in a network are using exactly the same measurement methods and reference materials

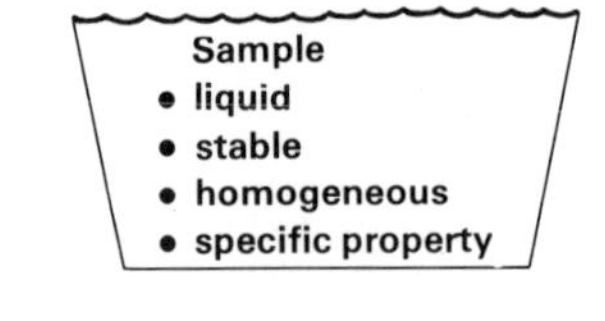

Lab A	Lab B	Lab N
$M_A \pm \sigma_A$	$M_B \pm \sigma_B$	$M_N \pm \sigma_N$

If all M's are accurate, then within the σ's

$$M_A = M_B = \ldots M_N$$

and

Measurement Compatibility Results

thus

ACCURACY ENSURES COMPATIBILITY

Figure 2. Schematic of how measurement accuracy assures measurement compatibility

value $\underline{M}$ within random uncertainties. If all the M's are accurate, then within the measurement uncertainties σ

$$M_a = M_b = \ldots M_n$$

and measurement compatibility results. Thus, again we emphasize that accuracy leads to compatibility.

The necessary requirements for accurate measurement systems are that they be both precise and free of systematic errors (7). It might also be desirable that accurate measurement techniques also have such characteristics as: wide dynamic range, rapid operation, and inexpensive operation. However, such requirements are not necessary to achieving accuracy but rather are practical considerations.

Consider a pragmatic operational definition of what we call "true value." The true value of a property is that value that can ultimately be traced to the base or derived units of measurement, e.g., length, mass, or time through experiments having no systematic errors (or with systematic error small relative to practical end-use requirements). One could look at the true value of the speed of light in two ways: 1. Philosophically--one may never be able to determine the true value with absolute certainty, and 2. Pragmatically--metrologists have been able to determine a best value of the speed of light with rather small uncertainties. This best value is operationally synonymous with the true value. Furthermore, these uncertainties in the measurement of the speed of light can be directly related to the uncertainties in the determination of the basic measurement units such as length or time. This paper assumes that the pragmatic approach to determining true values is valid. Thus, true values are those that are determined by precise measurement methods, that are essentially free of systematic error and are traceable to basic metrological measurements.

The Measurement Method Hierarchy and the Transfer of Accuracy

This leads us to the concept of the measurement method hierarchy, the problem of transferring accuracy throughout the hierarchy, and to the role of reference materials and reference methods in this process (8,9,10). There are a number of mechanisms that can be utilized to transfer accuracy throughout measurement networks. This is illustrated by referring to the so-called Measurement "Pyramid," which is shown inverted in figure 3.

At the bottom point are the fundamental metrologists of the world who are concerned with accurate experimental determinations of the base units of measurement. The metrological community consists of a relatively small group of scientists dedicated to the accurate determination of the seven basic measurement units such as length and the 40 or so derived units such as voltage. We must emphasize that unless such fundamental measurement experiments are carried out, the rest of the measurement infrastructure is on shaky grounds as far as accuracy is concerned. Metrological measurements are normally of the highest accuracy, in some cases approaching 1 part in 10^{10} or better.

The next level on the measurement pyramid is represented by absolute measurement methods (or definitive methods as they are being called in the clinical chemistry field). Definitive methods (11) are those that have been sufficiently well-tested and evaluated so that reported results have essentially zero systematic errors and have high levels of precision. Therefore they give true values within a narrow range of uncertainty. These methods are usually expensive, time-consuming, require highly sophisticated analysts and unfortunately are usually not very practical for everyday field use.

The third level is represented by other methods called by such names as reference methods or standardized methods. These, like absolute methods, are also of known accuracy although usually the inaccuracies are of greater magnitude or less well-defined than for definitive methods. These methods are generally faster and less expensive than the absolute methods. Many of the ASTM recommended analytical methods fall into this category (12).

The final level of the measurement pyramid consists of the routine or field methods that are generally fast, cheap, and usually require relatively non-sophisticated personnel for application. Field methods are generally used in applications involving large numbers of separate measurements that must be performed rapidly. Although many of the field methods may be very precise, they may be highly inaccurate.

This brings us to a key point of this paper. In order to achieve measurement compatibility on a large scale, some mechanism for transferring accuracy from the bottom two steps to the top of the pyramid must be found. One mechanism for transferring accuracy through this network is through well-characterized reference materials.

EXAMPLES OF THE TRANSFER OF ACCURACY

We now present some specific examples to illustrate the transfer of accuracy throughout the measurement hierarchy. At this point, we ignore the bottom step of the pyramid with the understanding that to achieve measurement accuracy, one must be able to trace certain key measurements to the experimental realization of the base measurement units.

The Measurement of Calcium in Serum

Consider the example of the measurement of Ca in serum, which is used by physicians to diagnose certain thyroid diseases. Figure 4 describes the measurement methods and associated inaccuracies at different levels in the calcium measurement hierarchy. The most accurate method for determining Ca is through the definitive method of isotope dilution-mass spectrometry (ID-MS) (13), whereby the accuracy can be evaluated from first principles and traced directly to the experimental realization of the base measurement units. As a reference material, a pure analyte such as $CaCO_3$ is required. The level of inaccuracy is 0.2% for calcium. The nationally-accepted reference method (14,15,16) is an atomic absorption technique. The accuracy of the reference method is based on a standard reference material which in turn was certified using the definitive method. The level of inaccuracy for the reference method is ±2.0%, a factor of ten higher than for the definitive method. Finally, there are numerous field methods, many of which have been assessed by the College of American Pathologists (CAP) in its proficiency testing program (17,18). In these tests, the CAP used the reference method with reference sera, which are "matrix" reference materials

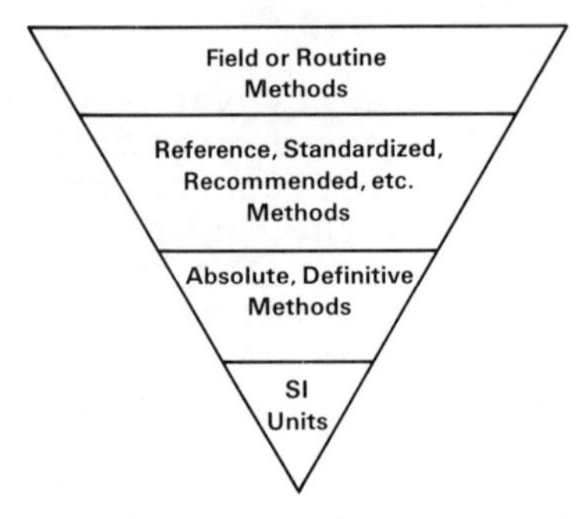

• Accuracy decreases and use increases in going from bottom to top.

Figure 3. The measurement "pyramid" representing the hierarchy of measurement methods needed to transfer accuracy from the experimental realization of the base measurement units to methods used in the field

	Method	Accuracy	Level	Reference Material
I.	Definitive Isotope-dilution Mass Spectrometry	via first principles; traceable to SI; pure analyte required (SRM)	±0.2%	Pure $CaCO_3$
II.	Reference Atomic Absorption	via samples accurately assayed using definitive method. SRM required.	±2.0%	Pure $CaCO_3$
III.	Field 9 different methods assessed by CDC	via reference method and accurately assessed ref-sera.	±5-10%	Ca in reference sera

Figure 4. The hierarchy of reference methods and reference materials used to establish a compatible measurement system for the determination of calcium in serum

containing known amounts of calcium in the matrix of interest. Depending on the specific field method, the inaccuracies are known to be in the range of 5-10%.

The recent efforts to improve the accuracy of calcium measurements in the United States illustrate a number of important points. These include the fact that an important component of a total national measurement system is some mechanism (e.g., proficiency testing) that assures long-term quality control of the system.

In 1969, NBS certified and issued for use in the clinical laboratories (SRM 915) $CaCO_3$. This event had very little immediate impact on the accurate measurement of calcium in serum on a national scale, because no nationally-accepted reference method was available at that time to evaluate the accuracy of the field methods. Recognizing this deficiency, NBS, together with several other government agencies [the Center for Disease Control (CDC) and the National Institutes of Health (NIH)] and several professional societies (e.g., the American Association for Clinical Chemistry), established a measurement network to develop a reference method for determining calcium in serum (14). Five "round-robin" tests using a network of seven qualified clinical laboratories were required before the accuracy of a reference method was sufficiently demonstrated. Reference method development is not a trivial undertaking: two years of effort costing over $125,000 were required in this case. An accuracy for the reference method of within ±1% of the true value was the initial goal. However, this level of accuracy simply was not attainable at that time and is now ±2%.

Having demonstrated this degree of overall accuracy for the reference method, a sample of serum was prepared and analyzed by seven participating laboratories (using the reference method and the SRM), and by nine commercial and hospital laboratories, which used field methods for calcium The resulting data are summarized in figure 5. All results are plotted as absolute percent deviations from the "true value" as determined by ID-MS. Note that three of the twelve commercial laboratories reported results above the 8% error danger line, which is the level where incorrect medical diagnosis could lead to erroneous treatment of the disease called hyperparathroidism. Obviously, some of the field methods were in need of improvement.

Subsequent to this study and based on the concept of using reference methods plus reference materials to help assure measurement accuracy and compatibility, the CDC (19) has discovered that of the nine different routine methods used for calcium, one is sufficiently biased that it is now recommended for discontinuance, while four others are in need of improvement. Field tests been used in the United Kingdom (20) parallel these findings.

During the development of the calcium reference method, the following factors were found to contribute to systematic biases: quality of reagents (including water); the quality of volumetric glassware; instrument stability and linearity; analytical techniques; the quality and even the motivation of the analyst. When factors such as these were properly identified and controlled, accurate and compatible measurement soon followed.

If measurement compatibility and accuracy on a national scale are to be maintained, these findings indicate the need for a mechanism to assure long-term quality control over the measurement process even in those cases where good methodology and standards have already been developed!

NBS has issued over 20 clinical reference materials in recent years and is currently involved in a joint program with CDC and the Food and Drug Administration to develop a number of clinical reference methods for substances such as glucose, various electrolytes in serum, urea and uric acid.

The Measurement of NO_2 in Ambient Air

An example of the use of SRM's to improve the accuracy of an important environmental measurement system, is the use of a series of NBS nitrogen dioxide SRM's to evaluate an official Environmental Protection Agency (EPA) reference method for the analysis of NO_2 in ambient air. This particular study is described in a paper by J. R. McNesby (21).

In 1971, EPA designated the "Jacobs-Hochheiser Method" to be the official EPA Reference Method for the measurement of NO_2 in ambient air. This is a colorimetric method involving the diazotization of sulfanilamide by ambient NO_2 in combination with other appropriate reagents. Because of legal necessity at the time to quickly designate an official Reference Method, it was not possible to evaluate the accuracy of the Jacobs-Hochheiser Method nor to perform any collaborative testing prior to its designation as an official Reference Method. This method was selected (22) in part because it had been previously used in a health effects study and was known to have good precision. Thus, in theory an internally consistent and compatible measurement system would result whereby unknown ambient air concentrations could be directly related to health effects data.

In 1972, EPA requested for and supported the issuance of a NO_2 Permeation Tube Standard Reference Material designated as SRM 1629 (23). This SRM consists of a glass tube filled with liquid NO_2. The NO_2 permeating through the cap is accurately measured via gravimetry. The tubes are calibrated for permeation rate as a function of temperature and are used to generate known concentrations of NO_2 in air.

In their first application, the SRM's were used by EPA (24) to evaluate the systematic errors inherent in the Jacobs-Hochheiser reference method. Previous applications of the Jacobs-Hochheiser method were based in part on the assumption that the collection efficiency was a constant 35% over the complete range of concentration. EPA used the SRM's to determine the collection efficiencies of the Jacobs-Hochheiser method and the results are shown in figure 6. Using four different permeation tubes, the overall collection efficiencies were found to vary considerably with NO_2 concentration. The 35% efficiency level is shown by the dashed line. The collection efficiencies at low concentrations were considerably higher than 35%. At high concentrations the collection efficiencies were considerably lower. These data showed the systematic errors inherent in the Jacobs-Hochheiser method. EPA withdrew the method as the official Reference Method for NO_2 measurements and has recently evaluated other more promising analytical methods. They have proposed that NBS Standard Reference Materials be used to calibrate the methods (25).

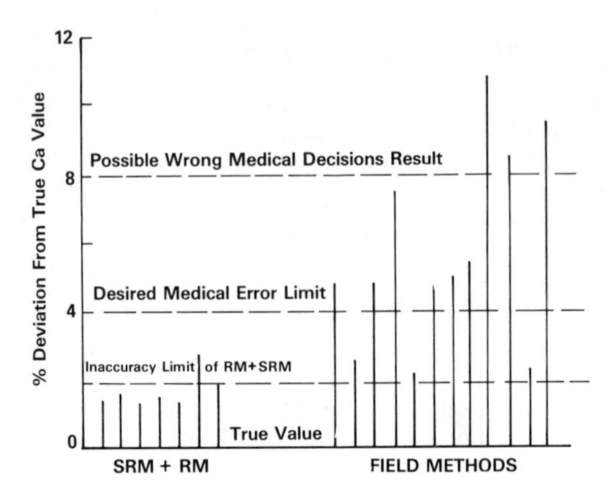

National Bureau of Standards Special Publications

Figure 5. A comparison of calcium in serum data obtained by laboratories using the SRM, plus reference method (RM) with results obtained by laboratories using field methods (14)

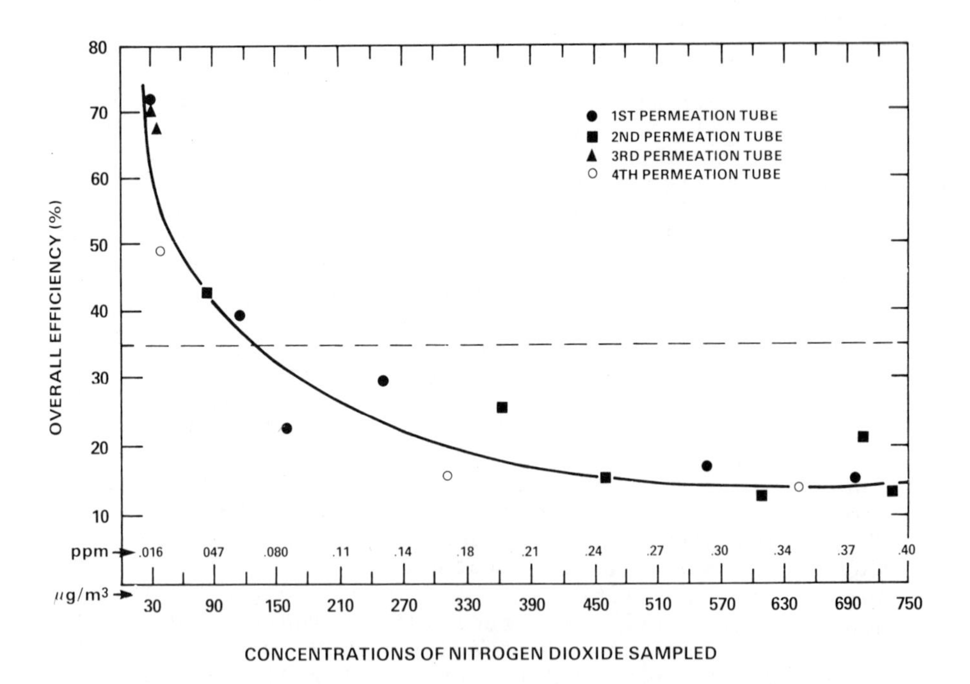

Figure 6. Data illustrating the use of NBS NO_2 permeation tube SRM's to evaluate the collection efficiencies of the Jacobs–Hochheiser method (21)

This example points out the great need for measurement methods with well-characterized accuracies as well as for good reference materials. The reference material by itself cannot assure interlaboratory measurement compatibility if the analytical methodology is poor. However, in such cases, the SRM provides an important means for evaluating analytical methods.

The Determination of Mercury in Water at the PPB Level

The need for good reference materials and analytical methods is illustrated again by the measurement of trace concentrations of mercury in water at levels of 1 part per billion or less. In early 1975, NBS was about ready to issue two new Standard Reference Materials to be used in low-level mercury analysis. SRM's 1641 and 1642 were certified for mercury composition at the level of 1.49 μg/ml and 1.18 ng/ml respectively. Since there were no mercury reference materials available at those levels, it was thought that the NBS-SRM's might provide an opportunity to evaluate analytical field methods currently being used for mercury monitoring. Here was a situation where there were no accepted national reference methods but there were supposedly adequate field methods (e.g., cold-vapor atomic absorption) in rather widespread use.

D. A. Becker (26) and coworkers in the NBS Analytical Chemistry Division carried out an interlaboratory comparison for mercury measurements. The participating laboratories performed experiments with and without SRM's to see if the SRM would help to improve measurement compatibility between laboratories. Seventeen laboratories participated in one or both phases of this study and eight laboratories submitted sufficient data to allow statistical analysis. Four different mercury concentrations were used to establish target values covering the range from 0.2 to 5.0 parts per billion. The raw data without the use of the SRM show a great deal of scatter with several of the laboratories obtaining results which deviated from the target values by a factor of 10 or more.

This experiment was repeated at the same concentration levels, but the NBS-SRM's were sent out along with the "unknown" solutions to be used as controls for checking the analytical procedures of the individual laboratories. The summary of results are shown in tables 1 and 2. Summarized are the target value, the average value obtained by all laboratories (with and without the use of the SRM) and the coefficient of variation between laboratories (with and without the use of the SRM). The data in table 1 indicate that the precision between laboratories was improved little if at all through the use of the SRM. Table 2 shows the deviations of the average values from the target value, expressed as a percentage of the target value with and without the use of the SRM. These results indicate little or no improvement in accuracy through the use of the SRM.

Thus, in this particular case the use of the SRM appears to have little or no effect on improving either the interlaboratory precision or the accuracy of the measurements. Any improvements are probably too small to be of practical value. The interlaboratory precision and accuracy seem to be much better above 1 ppb than below, with or without the use of the SRM.

The fact that the SRM does not help to significantly improve the accuracy or compatibility of these measurements leads us to examine the measurement methodology being used. A more detailed examination of the raw data indicates that only a few of the participating laboratories achieved good accuracy and precision with or without the SRM at all concentrations studied. For example, one particular laboratory deviated from the target value by only +11% and +4% without the SRM at the lowest and highest concentrations studied. Using the SRM, the deviations were +11% and +3% at the lowest and highest concentrations. Such data indicate that the measurement procedures of several of the laboratories were in very good statistical control. The data also provided some evidence that the laboratories coming closest to the target values also had greater within laboratory precision than the other laboratories.

It would be interesting to determine why certain laboratories were able to do so well while the measurements of others seemed completely out of control, even though a supposedly similar measurement technique was used by all. One approach to improving results in those laboratories would be to have the

Table 1. Average and Coefficient of Variation for Data Obtained by Eight Laboratories in the Evaluation of Methods Used to Determine Trace Mercury in Water

Target Value (ppb)	Without Use of SRM		With Use of SRM	
	Average(1)	% CV(2)	Average(1)	% CV(2)
0.18	0.24	57	.23	60
0.64	0.75	61	.58	49
1.16	1.30	25	1.22	16
4.98	5.17	7.3	5.12	9.3

(1) Average of 8 laboratories.

(2) Relative standard deviation, expressed as percent of the average (of 8 labs), of variability among laboratories.

Table 2. Deviation from the Target Values of the Interlaboratory Comparison Test Results for Mercury Measurements

	% Deviation From Target Value	
Target Value (ppb)	Without Use of SRM	With Use of SRM
0.18	+33	+28
0.64	+17	− 9
1.16	+12	+ 5
4.98	+ 4	+ 3

laboratories that are capable of producing results with good within-laboratory precision and accuracy, develop a strict measurement protocol or reference method. Then one could repeat the experiment requiring that the strict reference method protocol be followed by all laboratories. This should increase within laboratory precisions to the point where the SRM should then be of greater value in reducing between laboratory variations.

This example indicates that a reference material in and/of itself is not sufficient to insure accurate and compatible measurements, if a state of internal or within-lab quality control has not first been attained by each laboratory in the network. In addition to making proper use of the SRM, adequate analytical procedures and methods under strict quality control are also necessary.

Trace Cr in Biological Matrices

The final example is concerned with a lack of measurement compatibility between laboratories studying the role of trace metals in biological processes and systems. A recent paper by W. Mertz has reviewed the measurement problems associated with the analysis of the trace elements important to nutrition and health. For example, Cr is believed to play an important role in processes governing the onset of diabetes through a Cr-containing substance known as the "glucose tolerance factor."

Over the past few years, some serious problems associated with interlaboratory compatibility of Cr analyses in certain biological matrices have become apparent. Mertz reported on values for Cr concentrations in blood as obtained by various investigators using several different analytical methods (27). Variations by a factor of 1000 were reported. Even if one takes into account the fact that the investigators used different specimens and the differences one might expect in "normal" values between individuals, the trace chromium measurement system seems to be out of control.

This is further illustrated by some recent measurements performed at NBS and elsewhere (26). The Cr content of two biological matrix SRM's was determined. The NBS Orchard Leaves SRM (SRM 1571) has a known value for the Cr concentration, which seems to be easily reproduced when measured independently by different laboratories. On the other hand, when Cr is determined in the NBS Bovine Liver SRM (SRM 1577) at different laboratories, the values obtained show considerable variability. For example, NBS determined the total Cr content in Bovine Liver to be 170 ± 20 ppb using Neutron Activation Analysis with radiochemical separation. Another laboratory obtained 50 ppb using the same technique.

Not only is the total Cr content difficult to determine in Bovine Liver, but quantitative estimates of organic species of Cr are even more difficult to obtain. A common technique for measuring chromium is graphite furnace atomic absorption. However, some of the organic chromium associated with the glucose tolerance factor is apparently lost during the charring cycle--since it is volatile--thus leading to large measurement errors.

What we hope to do at NBS is to help resolve the Cr measurement problem by producing a Standard Reference Material with known certified concentrations of total Cr and also, hopefully, organic Cr. Brewers Yeast is the candidate biological matrix material. If such an SRM can be certified, it then can be used to optimize the atomic absorption techniques used to determine Cr and to help expedite the development of a reference method. The question of determination of speciation is particularly important because metallorganic Cr can be a factor of 100 more active in the glucose tolerance factor than inorganic chromium.

This example illustrates the fact that problems involving the quantitative measurement of distinct chemical species as opposed to elemental content are becoming increasingly important, particularly in the health and environmental areas. We believe that the accurate determination of the composition of distinct chemical species will present developers of reference methods and reference materials with a host of new and challenging measurement problems in future years.

CONCLUSION

Consider the problems in chromium analysis; in this case we are in the early stages of establishing a compatible national measurement system. One might

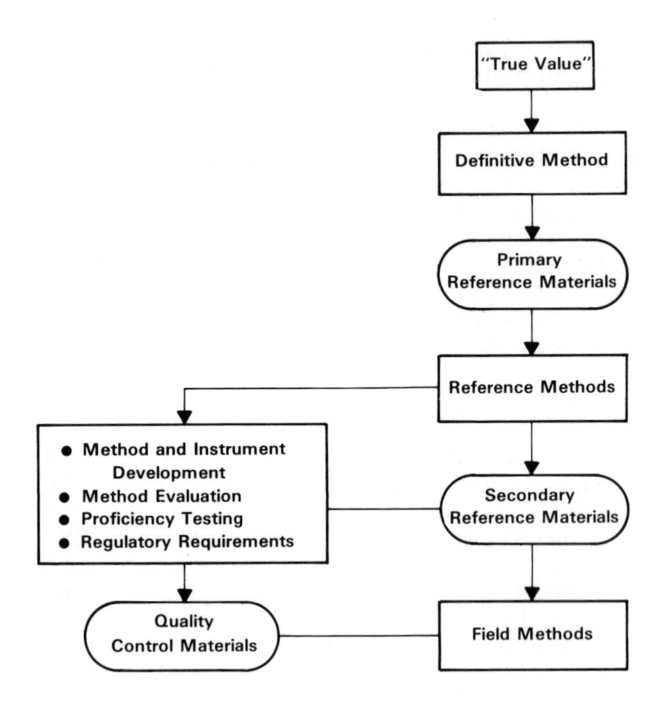

Figure 7. Schematic of an "idealized" measurement network needed to transfer accuracy and assure measurement compatibility

ask, what should be developed first--reference materials or reference methods? There is no definitive answer in general because each distinct type of measurement presents a unique set of problems and requirements. Good methodology (preferably definitive methods) is needed to certify SRM's and yet well characterized reference materials are obviously useful for evaluating methods (in particular reference methods).

This leads to the conclusion that the establishment of a compatible measurement system is an iterative, selfconsistent process sometimes requiring several generations of validating tests. This also leads to the concept of the idealized measurement network based on accuracy. There are a number of important mechanisms required to develop a compatible measurement system. They are: (1) An agreed upon system of base units of measurement; (2) Accurate reference methods and reference materials; (3) Field methods made compatible through the use of reference methods and reference materials. In addition, some type of a built-in feedback mechanism is needed for maintaining quality control of the whole system and for assuring compatibility between various components of the system.

One such idealized measurement system is shown schematically in figure 7. Measurement systems analogous to this are already well on their way to being implemented in environmental and clinical areas such as those previously cited. It is through networks such as these that measurement compatibility can be assured and maintained and that "meaningful" i.e., compatible measurements can be achieved on a national or international basis.

ACKNOWLEDGMENTS

The authors wish to acknowledge several helpful discussions with Dr. James R. DeVoe during the preparation of this paper. We are grateful to Dr. Donald A. Becker and coworkers for supplying us with the data on mercury and chromium measurements. We also thank Dr. James R. McNesby for supplying us with the data on the use of NBS NO_2 SRM's in evaluating the Jacobs-Hochheiser Method.

ABSTRACT

Reference Materials (called Standard Reference Materials SRM's by the National Bureau of Standards, NBS) are two important mechanisms being utilized to assure the accuracy and compatibility of measurements in large measurement systems. SRM's are materials whose properties (compositional and/or physical) have been well-characterized and certified by NBS. Reference Methods are analytical methods having high accuracy and precision, which have been thoroughly demonstrated. A systems approach to establishing accurate measurement systems is presented. Reference materials and reference methods assist in the transfer of accuracy gained in the experimental realization of base measurement units to the performance of measurements in the field. The application of the systems approach to "real world" situations is illustrated through the presentation of four examples: (1) The measurement of calcium in serum; (2) The determination of NO_2 in ambient air; (3) The analysis of trace levels of mercury in water; and (4) The measurement of chromium in biological matrices.

LITERATURE CITED

1. Cali, J. P., et al., "The Role of Standard Reference Materials in Measurement Systems," NBS Monograph 148, U.S. Government Printing Office, Washington, D.C. 20402 (1975).
2. Cali, J. P. and Stanley, C. L., Annual Review of Materials Science, 5, 329 (1975).
3. Cali, J. P., Med. Instru., 8, 17 (1974).
4. Wernimont, G., "Statistical Control of Measurement Processes," contained in this monograph.
5. Eilers, R. J., Clinical Chemistry, 21, 10 (1975).
6. Rhodes, R. C., "Importance of Sampling Errors in Chemical Analysis," contained in this monograph.
7. Eisenhart, C., Science, 160, (1968).
8. Seward, R. W., editor, "Standard Reference Materials and Meaningful Measurements," NBS Spec. Publ. 408, U.S. Government Printing Office, Washington, D.C. 20402 (1975).
9. Cali, J. P. and Reed, W. P., "The Role of NBS Standard Reference Materials in Accurate Trace Analysis," to be published in Accuracy in Trace

Analysis, NBS Spec. Publ. 422, U.S. Government Printing Office, Washington, D.C. 20402 (1976).
10. Cali, J. P., Bulletin of the World Health Organization, 48, 721 (1973).
11. Young, D. S., Z. Klin. Biochem., 12, 560 (1974).
12. See for example, 1973 Annual Book of ASTM Standards, parts 12, 32 and 42, published annually by the American Society for Testing and Materials, Philadelphia, Pa., 19103.
13. Moore, L. J., Anal. Chem., 44, 2291 (1972).
14. Cali, J. P., et al., "A Referee Method for the Determination of Calcium in Serum," NBS Spec. Publ. 260-36, U.S. Government Printing Office, Washington, D.C. 20402 (1972), reprinted (1976).
15. Cali, J. P., et al., Clin. Chem., 19, 1208 (1973).
16. National Committee for Clinical Laboratory Standards, NCCLS Proposed Standard: PSC-4 (1976).
17. Gilbert, R. K., Am. J. Clin. Pathol. 63, 974 (1975).
18. Hanson, D. J., Am. J. Clin. Pathol. 61, 916 (1974).
19. Private communication from J. Boutwell to J. P. Cali.
20. Pickup, J. F., et al., Clin. Chem. 21, 1416 (1975).
21. McNesby, J. R., Berichte Der Bunsen-Gesellschaft Fur Physicaliske Chemie, 78, 158 (1974).
22. "Air Quality Criteria for Nitrogen Dioxides," EPA, AP-84, (1971).
23. Hughes, E. E., "Development of Standard Reference Materials for Air Quality Measurements," International Instrumentation Automation Conference and Exhibit, ISA Reprint 74-704 (1974).
24. Federal Register, 38, 15174 (June 8, 1973).
25. Federal Register, 41, 11261 (March 17, 1976).
26. Becker, D. A., Private communication.
27. Mertz, W., Clinical Chemistry, 21, 408 (1975).

5

Optimization of Experimental Parameters in Chemical Analysis

STANLEY N. DEMING

Department of Chemistry, University of Houston, Houston, TX 77004

The title of this symposium, "Validation of the Measurement Process," may be interpreted in two ways:

In its usual usage, "validation" means "the determination of the degree of validity of a [measurement process]"(1). This definition suggests an activity that takes place after the measurement process has been developed. If the evaluation is successful, the process will receive official sanction, confirmation, or approval.

An alternate meaning of "validation" is "to make valid" in the sense of "producing the desired result" (2); that is, making the measurement process meet the criteria against which it is to be evaluated. This definition suggests activity that takes place while the measurement process is being developed.

This latter interpretation has been emphasized by Youden (3) and is the interpretation I wish to stress here if the initial development of a measurement process is carried out with the goal of meeting the evaluation criteria, then the probability that the process will receive rapid approval is greatly increased.

SYSTEMS THEORY

Figure 1 shows a systems theory view of the measurement process. The primary input to the system is a sample. The measurement process abstracts the desired information from the sample and transforms the information into a number. This number, or result, is the primary output from the system.

Ideally, the numerical value of the output should be related only to the desired information in the sample. As an example, a "perfect" measurement process for the analysis of enzyme activity would be sensitive to the amount of enzyme in the sample and insensitive to all other variables.

In practice, the numerical value of the output is influenced by a host of other factors. Some are associated with the sample matrix, while others appear as additional inputs to the measurement process. These factors may be systematized and are shown schematically in Figure 1.

An obvious categorization of the factors affecting a measurement process is the division into one set of factors that are known to have an effect on the process (solid arrows) and a second set of factors that do affect the results of the measurement process but have not yet been identified--that is, they are unknown (dashed arrows). Another grouping divides the factors into those that are controlled (represented by a dot on the tail of the arrow) and those that are uncontrolled. When factors are categorized in these two ways, four distinct types result:

A factor that is known to exert a significant influence on the result of a measurement process is usually controlled. This will usually improve the precision of the method if variations in the uncontrolled factor level appear as noise (that is, the variations are rapid with respect to the frequency of measurement); it might also improve the accuracy of the method if the frequency of calibration is long with respect to variations in the uncontrolled factor level.

Some factors are known to influence the result of a measurement process but are left uncontrolled. For example, if a factor is difficult or expensive to control and if the functional relationship of its influence is known, the level of this factor might be measured and a correction applied to the result. Or it might be known that a factor's influence on the result, though real, is not significant; it would probably be unnecessary to control such a factor.

Factors that are unknown and controlled are not usually a problem unless the method of control is inadvertently changed. A familiar example of a factor that is unknown and controlled is an impurity in a reagent: because the reagent is always added in a fixed amount, the level of impurity is also constant and is controlled. The effects of the reagent and the impurity are confounded and are usually not separated unless a change in reagent lot or supplier is made.

Most unknown factors are uncontrolled. It is assumed that factors in this category do not or will not exert a significant influence. Whatever influence they do exert is accepted as "noise" or imprecision.

The sample can contain factors from all of these categories (4).

RUGGEDNESS OF MEASUREMENT PROCESSES

As Mandel has pointed out, "The development of a method of measurement is to a large extent the discovery of the most important environmental factors and the setting of tolerances for the variation of each one of them" (4). Tolerances make possible the operational implementation of the concept of control: it is often impossible or impractical to control a factor *at* a given level, but it is usually possible and practical to control a factor *within* a specified domain of factor levels--that is, to control a factor around a given level, within specified tolerances. The specification of factor tolerances is based upon the required precision of the method and answers the question, "To what extent can a factor be allowed to vary before the output of the system changes by *y* amount?"

For a specified value of *y*, it is desirable that these tolerances be broad so that the measurement process is relatively insensitive to small variations in factor levels. To illustrate, consider the relationship between reaction rate (the result of a measurement process) as a function of pH (a known and controlled factor) for the kinetic determination of enzyme activity (see Figure 2). In general, enzymes do not function well at extremes of pH and exhibit an optimum with respect to pH. Let us assume that a method is to be developed for measuring the activity of an enzyme. A performance criterion has been spe-

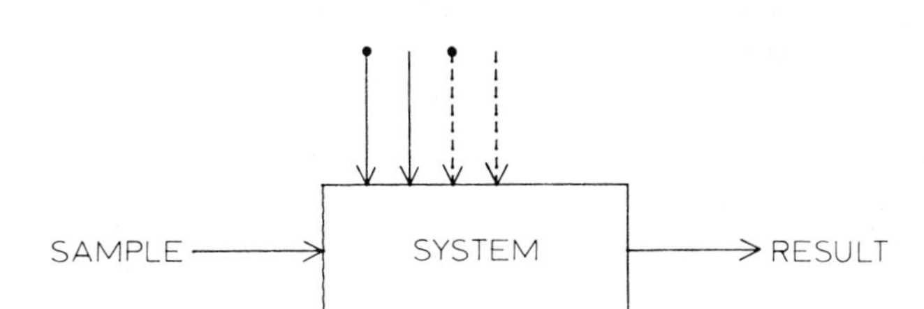

Figure 1. Systems theory view of the measurement process

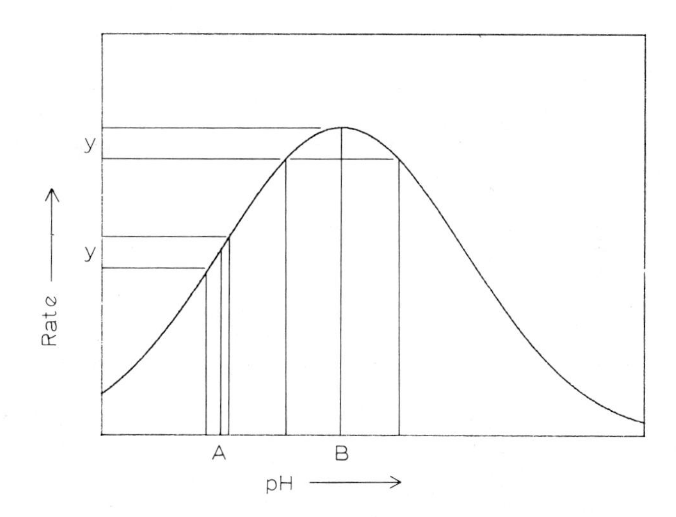

Figure 2. Reaction rate as a function of pH for the kinetic determination of enzyme activity

cified that requires an interlaboratory agreement within y reaction rate units.

If the initiating laboratory develops a method in which the pH level is set at point A in Figure 2, then the interlaboratory control of pH must be extremely tight: small differences in pH between the laboratories evaluating the method will show up as a large between-laboratory variance. Worse still, in the absence of additional information, it would be difficult to separate pH as one of the causes of this variance.

If, instead, the initiating laboratory suggests a method in which the pH level is set at point B in Figure 2, then small differences in pH between the laboratories evaluating the method will contribute very little to the between-laboratory variance. This method would have a higher probability of being accepted after its first interlaboratory test.

Many of the factors affecting measurement processes exhibit the behavior shown in Figure 2. Other factors initially increase and then asymptotically approach a plateau (e.g., the substrate dependence of many enzymes). With factors exhibiting these types of behavior, adjusting the factor levels to improve the system output will also improve the factor tolerances (5).

DEVELOPMENT OF MEASUREMENT PROCESSES

The development of a measurement process should involve three stages: obtaining a response, improving the response, and understanding the response.

Many laboratories carry the development through the first stage only. Youden (3) has pointed out the potential limitations of such methods and has emphasized the importance of acquiring an operational understanding of the measurement processes; that is, identifying and controlling those factors that exert a significant effect on the system. This is especially critical if the measurement processes are to become widely used by a number of laboratories.

The improvement of response has been carried out infrequently, although its importance has been recognized for some time. In 1952, Box (6) presented a paper in which it was pointed out that single-factor-at-a-time strategies are inadequate for optimizing most chemical processes (7,8) and that measurement processes (conceptually very similar to production processes) can be effectively improved by the use of sequential factorial designs, a technique that later became known as "evolutionary operation," or EVOP (9, 10). EVOP strategies are well suited to the industrial environment--the production process is being run constantly and provides a continuous framework for the large number of experiments required by the sequential factorial designs. In the developmental laboratory, however, efficiency of initial experimentation is stressed and EVOP strategies are less desirable. In 1962, Spendley, Hext, and Himsworth (11) introduced the fixed size simplex as a more efficient sequential experimental design for traditional evolutionary operations; Long (12) appears to have been the first to apply fixed size sequential simplex designs to the development of measurement processes. Nelder and Mead (13) modified the sequential simplex method to allow acceleration in directions that are favorable and deceleration in directions that are unfavorable.

We have found the variable size simplex (slightly modified) to be a rapid means of improving results in the development of analytical chemical measurement processes (14-17). Factorial designs (18) , central composite designs (19), and Box-Behnken designs (20) are useful for understanding the various factor effects upon the response in the region of the optimum.

EXAMPLE

The determination of formaldehyde in an aqueous sample can be determined by the addition of chromotopic acid (4,5-dihydroxy-2,7-naphthalendisulfonic acid) and sulfuric acid (21-25); a color develops, and the absorbance is read at 570 nm.

In this study (14), a sample size of 2.00 ml was chosen. The amount of aqueous 20 g l^{-1} chromotropic acid (CTA, factor x_1) was allowed to vary between 0.00 and 1.00 ml; concentrated sulfuric acid (H_2SO_4, factor x_2) could vary between 1.00 and 10.00 ml. The objectives of the study were: (a) to determine the amounts of H_2SO_4 and CTA that produced the greatest absorbance for a given amount of formaldehyde (2 ppm)

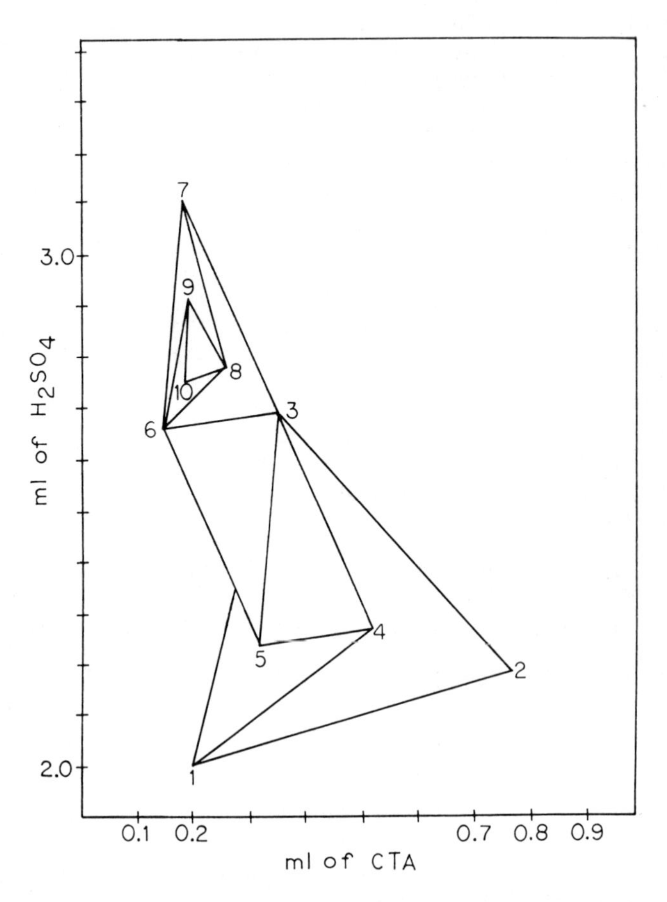

Analytica Chimica Acta

Figure 3. Simplex progress in the chromotropic acid-concentrated sulfuric acid domain. See text and Table 1 for details (14).

TABLE 1

Simplex progress

Vertex	Retained vertices	Chromotropic acid (ml)	Sulfuric acid (ml)	Absorbance
1	-	0.200	2.00	0.221
2	-	0.780	2.18	0.080
3	1,2	0.355	2.68	0.531
4'[a]	-	-0.224	2.50	-1.000[b]
4	1,3	0.529	2.26	0.223
5'	-	0.684	2.94	0.197
5	3,4	0.321	2.23	0.325
6	3,5	0.147	2.65	0.563
6'	-	-0.044	2.85	-1.000[b]
7	3,6	0.182	3.09	0.562
8'	-	-0.027	3.07	-1.000[b]
8	6,7	0.260	2.77	0.573
9'	-	0.226	2.33	0.502
9	6,8	0.193	2.90	0.599
10'	-	0.305	3.03	0.570
11	8,9	0.187	2.74	0.584

[a]Primes indicate rejected vertices.

[b]Boundary violation.

Reprinted from reference 14 with permission of Elsevier Scientific Publishing Company.

TABLE 2

Results of factorial experiments

Chromotropic acid (ml)	Sulfuric acid (ml)	Absorbance
0.10	2.50	0.524
		0.538
	2.80	0.515
		0.516
	3.10	0.526
		0.530
0.30	2.50	0.455
		0.509
	2.80	0.583
		0.575
	3.10	0.534
		0.545
0.50	2.50	0.386
		0.428
	2.80	0.545
		0.537
	3.10	0.554
		0.551

Reprinted in part from reference 14 with permission of Elsevier Scientific Publishing Company.

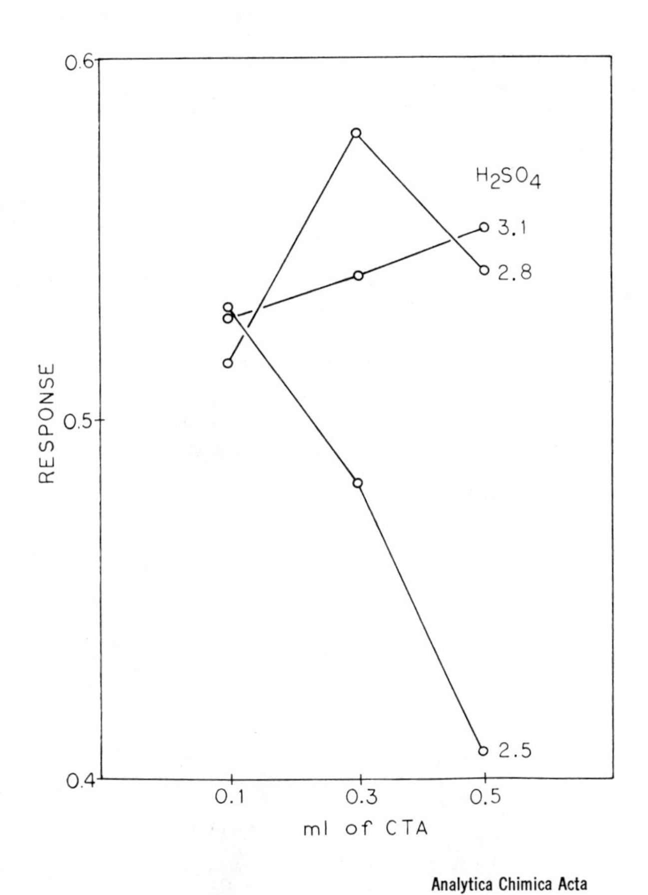

Analytica Chimica Acta

Figure 4. Cell mean plot of factorial study (14)

TABLE 3

Factorial analysis of variance (ANOVA)

Source of variation	Degrees of freedom	Sum of squares	Mean square	F-ratio	Significance (%)
Chromotropic acid	2	0.00369	0.00179	6.27	98.0
Sulfuric acid	2	0.01926	0.00963	33.66	99.9
Interaction	4	0.01670	0.00417	14.59	99.9
Error	9	0.00258	0.00029	-	-

Reprinted in part from reference 14 with permission of Elsevier Scientific Publishing Company.

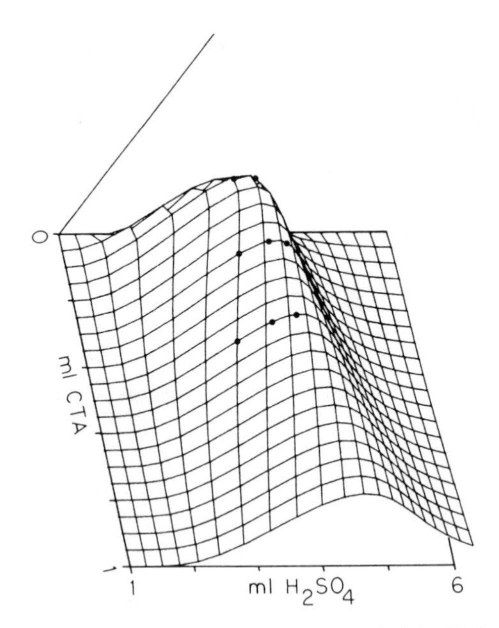

Figure 5. Absorbance response surface as a function of chromotropic acid volume, and concentrated sulfuric acid volume (14)

and (b) to understand the effects of H_2SO_4 and CTA upon the response in the region of the optimum so that factor tolerances could be specified.

Figure 3 shows the progress of the simplex toward the optimum (data in Table 1). The numbers in the figure indicate the sequence in which the retained vertices were evaluated; rejected vertices are not shown. Table 2 contains the results of a three-level two-factor, full factorial study with replication carried out in the region of the simplex optimum; the results of the analysis of variance are presented in Table 3. In Figure 4, cell means are plotted *vs.* CTA for each of the three levels of H_2SO_4.

Other studies (25) have shown that the absorbance response is related to the ratio (sulfuric acid volume)/(total volume); this is probably an effect caused by the heat of mixing which drives the reaction toward completion. Assuming this relationship to be approximately Gaussian in the region of the optimum, a model of the form

$$\text{Absorbance} = k_1 \exp[-(((x_2/(x_1+x_2+2.0))-k_2)^2)/(2k_3^2)] \\ *(2.0/(x_1+x_2+2.0))(1-\exp(-k_4 x_1))$$

can be fit and is visualized in the pseudo-three-dimensional plot shown in Figure 5. The factorial points are superimposed on the surface.

At low H_2SO_4 volumes, increasing the volume of CTA moves across the "front" of the response surface with the result that response decreases. At an intermediate level of H_2SO_4, increasing the volume of CTA moves from "behind" the diagonal ridge to the top of it and down again on the front side. At the highest level of H_2SO_4 studied, increasing the volume of CTA moves along the "back" of the ridge with the result that response increases.

With this operational understanding of a process for measuring the amount of formaldehyde in an aqueous sample, factor levels and factor tolerances can be specified. Because concentrated H_2SO_4 is a worrysome reagent, and because aqueous CTA solutions of accurate concentration are easily prepared and handled, it would be appropriate to specify a tight CTA level of 0.1 ml where the volume of H_2SO_4 has less of an effect.

CONCLUSION

The strategy of obtaining a response, improving the response, and understanding the response is a reasonable means of obtaining sound measurement processes. We have found the variable size sequential simplex design to be an efficient means of optimizing the primary response from a measurement process. Established statistical designs allow an understanding of the factor effects and their interactions in the region of the optimum.

ACKNOWLEDGMENTS

The following have contributed to the ideas and work presented here: P. G. King, S. L. Morgan, L. R. Parker, Jr., A. S. Olansky, and L. A. Yarbro. The author acknowledges support from the National Science Foundation through grants GP-32911 and MPS-74-23157.

LITERATURE CITED

1. "Webster's New Collegiate Dictionary," 1292, G. & C. Merriam Company, Springfield, MA, 1973.
2. "The Random House Dictionary of the English Language," 1578, Random House, New York, NY, 1969.
3. Youden, W. J., Materials Research & Standards (1961), 1, 862.
4. Mandel, J., "The Statistical Analysis of Experimental Data," Interscience, New York, NY, 1964.
5. Skogerboe, R. K., in Baer, W. K., Perkins, A. J. and Grove, E. L., Eds., "Developments in Applied Spectroscopy," Vol. 6, 127, Plenum, New York, NY 1968.
6. Box, G. E. P., Analyst (1952), 77, 879.
7. Box, G. E. P., Biometrics (1954), 10, 16.
8. Morgan, S. L., and Deming, S. N., Anal. Chem. (1974), 46, 1170.
9. Box, G. E. P., Appl. Statist. (1957), 6, 81.
10. Box, G. E. P., and Draper, N. R., "Evolutionary Operation," Wiley, New York, NY, 1969.
11. Spendley, W., Hext, G. R., and Himsworth, F. R., Technometrics (1962), 4, 441.
12. Long, D. E., Anal. Chim. Acta (1969), 46, 193.
13. Nelder, J. A., and Mead, R., Computer J. (1965), 7, 308.

14. Olansky, A. S., and Deming, S. N., Anal. Chim. Acta (1976), 83, 241.
15. Morgan, S. L., and Deming, S. N., J. Chromatogr. (1975), 112, 267.
16. Parker, L. R., Morgan, S. L., and Deming, S. N., Appl. Spectrosc. (1975), 29, 429.
17. Deming, S. N., and King, P. G., Research/Development (1974), 25(5), 22.
18. Fisher, R. A., "The Design of Experiments," Oliver and Boyd, Edinburgh, 1935.
19. Box, G. E. P., and Wilson, K. B., J. Royal Statist. Soc., B (1951), 13, 1.
20. Box, G. E. P., and Behnkey, D. W., Ann. Math. Statist. (1960), 31, 838.
21. Bricker, C. E., and Vail, W. A., Anal. Chem. (1950), 22, 720.
22. Klein, B., and Weissman, M., Anal. Chem. (1953), 25, 771.
23. Kamel, M., and Wizinger, R., Helv. Chim. Acta (1960), 43, 594.
24. Sawicki, E., Hauser, T. R., and McPherson, S., Anal. Chem. (1962), 34, 1460.
25. Houle, M. J., and Powell, R. L., Anal. Biochem. (1965), 13, 562.

6

Components of Variation in Chemical Analysis

RAYMOND C. RHODES

Environmental Protection Agency, Research Triangle Park, NC

The first task in the evaluation of any analytical chemistry method is to determine the sources of variability of the results and then to minimize the total variability by searching out and controlling the major contributors. There are many sources of variability in a chemical analysis process. This work describes several of these sources that have been encountered in work by the Environmental Monitoring and Support Laboratory of EPA.

Variability can be expected to occur in any or all of the following measurement steps within a given laboratory:

1. The material to be analyzed
2. Materials, including reagents, used in the analysis
3. Calibration materials or devices
4. Environmental factors
5. Analysts
6. Instruments, or apparatus

While these items are not definitive, they describe the general classes of sources of variability that must be considered. It is clear that when one is measuring the reproducibility of an analytical chemistry method, it is important that the complete method's variability is being measured. For example, in the case where events are measured under the Poisson probability distribution it is usually improper to describe the error as resulting from this effect alone. In another example, the replication of only a portion of the measurement method, such as replicate analysis of a single extraction, is sometimes erroneously quoted as the precision of the method.

Every laboratory should systematically maintain a compilation or record of the effects of each of the above-listed factors, as obtained from special studies conducted in their laboratory or the laboratories of others, so that if improvement is needed in the quality of the reported results, efforts to effect such improvements can be made in the most cost-effective way. The planning of such special studies and the analysis of the data therefrom is an area where statisticians and chemists, working together, can gain much in the knowledge and understanding of the contributions to the total measurement variability. Each laboratory should conduct periodic quality control checks to measure and control the combined effects of the sources of variability which affect their reported results.

THE LABORATORY MEASUREMENT PROCESS

A schematic diagram of portions of the measurement process of an industrial laboratory is presented in Figure 1. Inputs of physical samples to the laboratory are shown:

(1) Samples from the manufacturing process

(a) Raw materials
(b) In-process materials
(c) Final product

(2) Calibration standards
(3) Reagents and other materials

The various internal factors of the laboratory measurement process, previously mentioned, are also shown.

The measurement process and sampling considerations for pollutant measurements are identical except for different types of sampled materials to be analyzed. The schematic of Figure 1 generally applies to any measurement process, whether of industrial, research, government or independent laboratories.

Several recent articles and documents discuss the various quality assurance aspects of the laboratory measurement process. The measurement or analytical process in industrial situations is normally a part of the overall quality assurance system. Moreover, in addition to being a part of the quality assurance system, recent concepts, quite appropriately view the measurement or analytical process as one which should maintain its own quality assurance sub-system or program (1, 2, 3). In the Environmental Protection Agency (EPA) the environmental monitoring efforts may be viewed as a production process which is essentially a measurement process, having as its product, environmental DATA. Accordingly, EPA is issuing guidelines for Quality Assurance Programs for each of its air measurement methods (4), and has issued a "Quality Assurance Handbook for Air Pollution Measurement Systems" (5). The above concept is recognized in EPA as not only appropriate for monitoring programs, but also for research projects in which it is desired to obtain data of high quality. Major research projects require rather extensive quality assurance programs (6).

The accuracy of the analytical portion of a chemical measurement system is achieved by the use of Standard Reference Materials, high quality reagents, and by control of variables of the calibration process. It should be emphasized that the basis for accuracy of the results from a laboratory should emanate from outside the laboratory. Unless a laboratory has its own internal capability for producing primary standards, it must depend upon some external source of standards. Otherwise, it is attempting "to lift itself by its own bootstraps." If unquestioned accuracy is desired, all measurements should be traceable to the standards of the National Bureau of Standards or other national and international standards laboratories. Even with traceable standards, accuracy depends upon their proper care and use. The certificates issued by the National Bureau of Standards for its Standard Reference Materials (SRM) often include carefully-worded cautions which explain that the SRM's have specific expiration dates and that they must be given proper care, and used under specified conditions. In fact, the accuracy of calibrations depends not only upon the standard used, but upon the entire calibration process (7).

The relative accuracy among laboratories can be measured through the use of interlaboratory comparison studies, represented by an input-output relationship on Figure 1. The objective of interlaboratory comparison studies, in which the same samples are analyzed by various laboratories, is to make comparisons with respect to accuracy. Two types of interlaboratory studies are conducted. Collaborative studies are generally considered as those studies which are conducted among a group of selected laboratories to evaluate a newly developed analytical method (8, 9). Other interlaboratory studies are conducted to compare the results of standard methods in use at various selected or volunteering laboratories (10). In many cases interlaboratory studies of either type reveal startling differences in accuracy between laboratories even though special attention and care (not routinely used for internal measurements) may have been taken. Such results support the fact that accuracy depends upon many factors other than standards. The occurrence of excessively deviant results from a laboratory participating in an interlaboratory study should trigger the initiation of investigational efforts to determine the cause.

As described in Chapter I, controls for precision may be considered at two levels, i.e., "local" control and "regional" control.

CONTROLS FOR PRECISION

Local Control	Regional Control
Duplicates, Back-to-Back	Duplicates by
Duplicates, Same Run	Different Analysts
Duplicates, Same Day	Different Equipment
	Different Calibrations
	Different Days

For a given method, there are usually a number of duplicate or replicate measurements of the same sample which can be made to control and measure the precision of the partial or complete measurement process. Sequential duplicates for a particular part of the measurement process are the most "local" of controls. They are useful in assuring control by the particular combination of sample/analyst/equipment/-procedure/conditions/time, and thus should show the greatest possible precision. As other normally existing variables of the measurement process

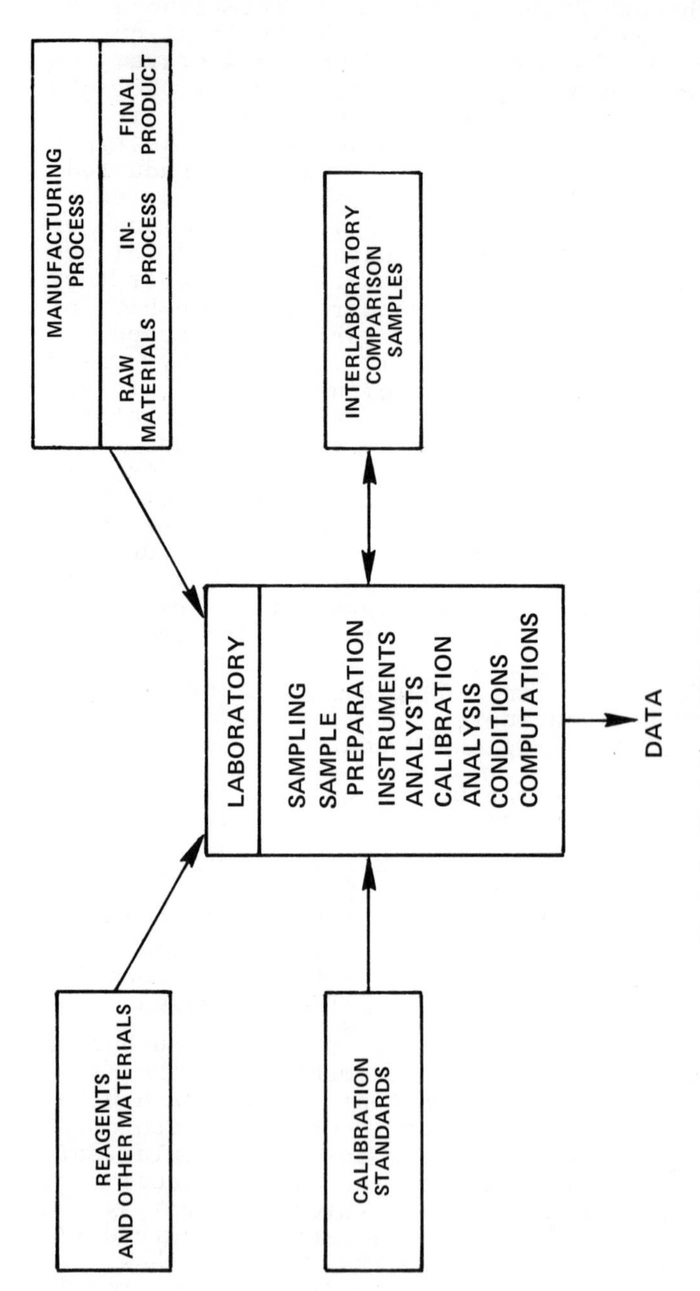

Figure 1. The industrial laboratory measurement process

(different analysts, different equipment, different calibrations) enter into the replications, lower precision results. These can be considered to be of a "regional" control nature, and always exist in the data for between-laboratory comparisons but normally are not reflected in most within-laboratory precision estimates. Nevertheless, they are part of the within-laboratory total measurement variability, and require special studies to ascertain their effects.

There are many levels of precision and it is very important that the specific variables which have entered into the comparison be identified (11, 12). If in the normal course of events a given sample may have been analyzed by any one of several instruments, with one of several calibrations, by any one of a number of analysts, and at one of several times under any of several conditions, then the measure of precision relating to the results for that particular sample should involve all of the likely variables. Therefore, a most realistic measure of precision for a given method at a given laboratory would be one obtained by having had all routinely existing variables 'at play' during the determination. Obviously, it is idealistic and impractical to conduct routine analyses under all likely combinations. However, such precision estimates should be determined by combining all available information on individual precision estimates. It may be necessary to conduct some special studies to fill any gaps. And such overall precision estimates should be periodically re-evaluated, particularly if changes are introduced into the measurement system.

Changes are frequently introduced in measurement systems either because of undesired necessity or internationally desired improvement. Any change may affect the accuracy and/or precision of the measurement system either in a desirable or undesirable manner. Statisticians and quality assurance people are particularly suspicious of change. Many times significant undesirable shifts in the level of analytical results have been determined to have been caused by the coincidental introduction of seemingly insignificant change in procedures, equipment, or materials. The slogans of statisticians and quality assurance people might well be "Cave Vicissitudines" (beware of changes) or "Cave Varietas" (beware of differences).

A very important but yet very simple principle, which should be used whenever changes are introduced into the measurement process is "The Overlap Principle." For example, overlapping some "old" variable, A, with a "new" variable, B, should be employed whenever practicable to obtain objective data by comparing a "before" condition with the "after" condition to assure that the change has not introduced a significant or deleterious effect. For example, if a new instrument, new analyst, new reagent source (supplier), or even a new calibration source is introduced into the measurement process, it is a good quality control procedure to analyze a given sample or samples of material under both the "old" and the "new" conditions, to assure that the change has not introduced some undesirable deleterious effect.

SAMPLING AND ANALYSIS VARIABILITY OF EPA REFERENCE SAMPLES

An example from an environmental measurement laboratory is used to illustrate the determination of the effects of several sources of variability in a measurement system.

The Environmental Monitoring and Support Laboratory periodically distributes to various environmental laboratories in the United States, carefully prepared reference samples as a check on the accuracy of the results from the laboratories. Although the samples are prepared as nearly identical as possible, some variability between the samples cannot be completely eliminated. To assess the average concentration level of the samples and to determine the sample-to-sample variability, EPA analyzes a number of randomly selected samples prior to distributing the remainder to the various participating laboratories. In the example to follow, EPA analyzed each sample on two different days to obtain some measure of between-day variability. Although there are a number of sources of variability in the entire measurement system, only the following four will be evaluated, namely:

1. Sample-to-Sample variability
2. Within-sample Within-day variability
3. Day-to-Day variability
4. Laboratory-to-Laboratory variability

The measurement method involved in the example is the pararosaniline method for analysis of sulfur dioxide (13). As part of EPA's quality assurance program, samples of sodium sulfite in tetrachloromercurate solution are carefully prepared to simulate field samples, and these samples are distributed to the various participating laboratories for analysis.

A large number of samples at each of five different concentrations (or series) are prepared by a contractor. Each sample, consisting of one-half milliliter of solution, is freeze-dried and then sealed in a glass vial to maintain the integrity of the sample until it is ready for analysis by the participating laboratory. EPA analyzed a small random sample from each of the five concentration levels to estimate the variation among samples. In the analysis, the material in the vial is dissolved and diluted to a total volume of 50 milliliters with 10 milliliter aliquots taken on each of two days for analysis.

For brevity, only the results for Series 1000 (the lowest concentration level) and Series 5000 (the highest concentration level) will be considered in detail. The results of the individual analyses are presented in Table I.

Statistical analysis of variance technique was used to estimate for each series,

1. The between-sample variability, σ_s
2. The between-day variability, σ_d
3. The within-sample, within-day variability σ_e

As an example, the results of the analysis of variance for series 5000 are shown in Table II.

From the analysis of variance given above

$$.3516 = \hat{\sigma}_e^2 + d\hat{\sigma}_s^2$$

$$5.7608 = \hat{\sigma}_e^2 + s\hat{\sigma}_d^2$$

$$.1034 = \hat{\sigma}_e^2$$

where d = number of days
s = number of samples

Table I. Individual Results (in ug) of SO_2

Series 1000

Sample No.		Day 1	Day 2	Difference
1381		11.24	11.36	- .12
1427		9.84	11.01	-1.17
1436		10.54	11.01	- .47
1376		10.54	11.01	- .47
1490		11.24	11.01	.23
	Average	10.680	11.080	

Series 5000

Sample No.		Day 1	Day 2	Difference
5589		54.07	56.22	-2.15
5750		54.07	55.86	-1.79
5649		55.12	56.22	-1.10
5557		55.47	56.57	-1.10
5401		54.77	56.22	-1.45
	Average	54.700	56.220	

Table II. Analysis of Variance

Source	Sum of Squares	Degrees of Freedom	Mean Square	F-test	Estimated Mean Square
Between-Sample	1.40634	4	.3516	3.40	$\sigma_e^2 + d\sigma_s^2$
Between-Days	5.76081	1	5.7608	55.69	$\sigma_e^2 + s\sigma_d^2$
Within Sample-Day	.41374	4	.1034		σ_e^2
Total	7.58089	9			

Solving for $\hat{\sigma}_s^2$ and $\hat{\sigma}_d^2$ gives

$$S_s = \hat{\sigma}_s = .352$$

$$S_d = \hat{\sigma}_d = 1.064$$

$$S_e = \hat{\sigma}_e = .322$$

The above example is presented to illustrate the basic principles in determining the components of variance. Most often the experimental design for the analysis of components of variance is more complicated than this (14); however this fact, which usually requires a statistician's assistance in planning such a study and in analyzing the data therefrom, does not detract from the importance of this type of study.

The components of variation resulting from the five separate analyses are presented in Table III.

A general pattern should exist across levels for the three components of variation. The standard deviations for the within sample-day variability and between sample variability are plotted on Figure 2 for the various concentration levels. Since no general pattern, either increasing or decreasing, exists, the pooled, i.e., statistically averaged, values for these standard deviations are:

s_e = .276 for within sample-day

s_s = .394 for between-sample

The plot of the between-day standard deviations (Figure 3), however, does show a general increasing pattern with increasing concentrations, with the exception of series 2000. Omitting the results for series 2000, the standard regression relationship of the between-day standard deviation to concentration level is as shown on Figure 3. The statistically more correct weighted* regression relationship is also shown, and is appreciably different from the standard

---

*A weighted regression is appropriate because of violation of the assumption of homogeneous variances--the variation of sample standard deviations is greater for larger expected or true standard deviations.

Table III. Components of Variance Obtained from Analysis of Variance

Series	Within Sample-day (S_e)	Between-Sample (S_s)	Between-Day (S_d)
	Standard Deviations, μg		
1000	.367	.221	.230
2000	.207	.493	zero*
3000	.285	.401	.714
4000	.136	.447	.804
5000	.322	.352	1.064
Pooled	.276	.394	

*(The zero value--actually it was negative--is possibly due to peculiar combinations of individual values. In this particular study, the lack of a between-day effect for series 2000 is suspected to be due to some assignable but unknown cause.)

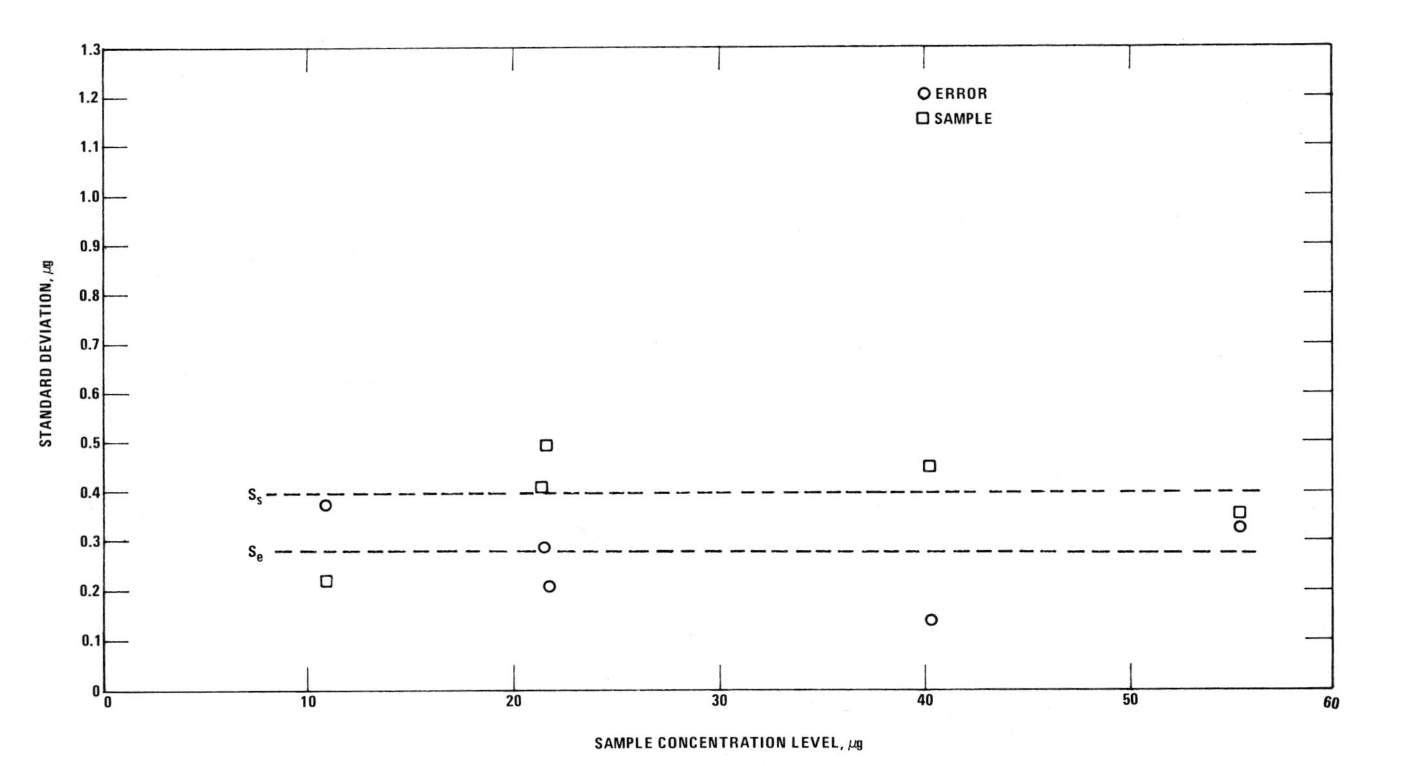

Figure 2. Within sample-day (error) variability; between sample (sample) variability vs. sample concentration levels

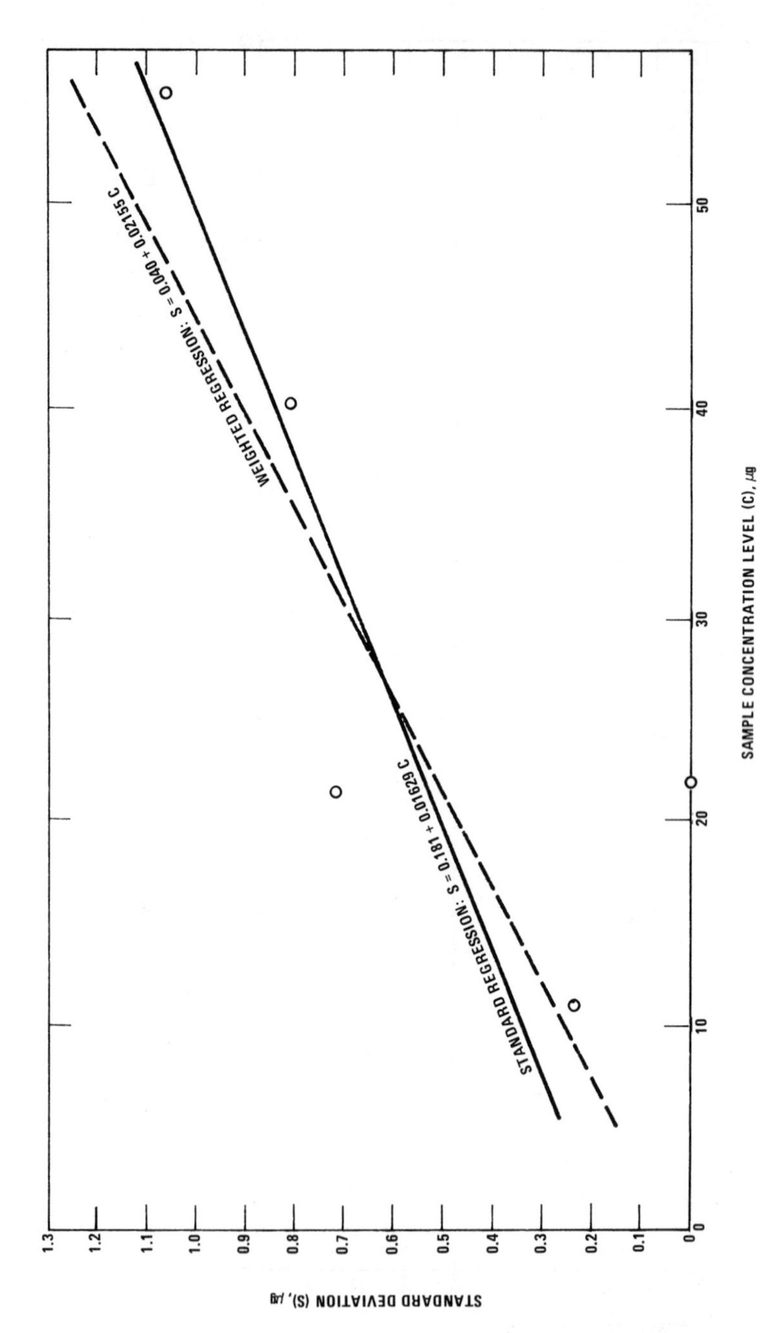

Figure 3. Between day variability vs. sample concentration levels

regression relationship. From the weighted regression line the expected between-day standard deviations are as follows:

Series	Standard Deviation, μg
1000	.275
2000	.511
3000	.504
4000	.911
5000	1.235

The above results were used to determine appropriate acceptability limits for the laboratories participating in an interlaboratory survey when analyzing the remaining samples of the five series. It is assumed that other laboratories receive only one sample vial from each of the five series and that they analyze each sample only once on some given day. (This does not mean that they have to analyze more than one sample on any given day.) It is also obviously assumed (1) that the samples analyzed by the EPA are representative random samples from the total of each series of sample vials, (2) that the two days during which EPA analyzed the samples are representative of any day on which an analysis may be performed, and (3) that the participating laboratories have capability for accurate and precise analyses at least equal to that of EPA.

Table IV shows the determined standard deviations for:

1. within sample-day variability,
2. sample variability,
3. day-to-day variability, and
4. combined variability

for each series.

In computing the acceptability limits for the interlaboratory comparisons, two variabilities must be considered.

1. The variation of the estimate of the "true" level from EPA analysis.

2. The combined within sample-day, between-sample, and between-day variation for the

Table IV. Combination of Variation to Determine Acceptability Limits for Interlaboratory Comparisons

	Standard Deviations							
Series	Within Sample-Day	Between-Sample	Between-Day	Combined S_t	S_{true}	95% Acceptability Limits	Upper Limit	Lower Limit
1000	.276	.394	.275	.554	.277	10.880 ± 1.239	12.119	9.641
2000	.276	.394	.511	.702	.411	21.864 ± 1.627	23.491	20.231
3000	.276	.394	.504	.697	.407	21.513 ± 1.614	23.127	19.899
4000	.276	.394	.911	1.030	.674	40.388 ± 2.462	42.850	37.926
5000	.276	.394	1.235	1.325	.895	55.459 ± 3.198	58.657	52.261

single result obtained by a participating laboratory.

The standard deviation which is used as a basis for computing the acceptability limits of Table IV is the statistical combination of 1 and 2, as follows:

$$s_{total} = s_{\text{"true"}}^{2} + s_{c}^{2}$$

where $s_{\text{"true"}}$ = standard deviation of the estimated "true" concentration level determined by EPA

s_c^2 = the combined within sample-day, sample, and day variability of a single result from a participating laboratory

$= s_s^2 + s_d^2 + s_e^2$

s_s = standard deviation of the sample variability

s_d = standard deviation of the day-to-day variability

It has been shown by statistical theory that when combining the errors of independent variables, the squares of the standard deviations, i.e., the variances, are additive. (In some fields, the above addition of the variances is referred to as the Root-Mean Square or RMS sum.)

The variability of the estimated "true" concentration level for each series is given by the following expression:*

$$s_{\text{"true"}} = \frac{s_s^2}{5} + \frac{s_d^2}{2} + \frac{s_e^2}{10}$$

---

*If necessary, consult your local statistician for the basis of this expression.

These calculated values are:

Series	$s_{\text{"true"}}$
1000	.277
2000	.411
3000	.407
4000	.674
5000	.895

For each series, the total standard deviation for expected differences between the EPA value and a participating laboratory result is as shown in the following table:

Series	s_{total}
1000	.620
2000	.814
3000	.807
4000	1.231
5000	1.599

The corresponding 95% acceptability limits are therefore:

$$\bar{x} \pm 2s$$

where $\bar{x}$ is the average of the 10 values of each series.

The computed values are shown in Table IV. It should be emphasized that these limits are minimum limits. Based on the information in the data reported herein, agreement of the participating laboratories' results with those of EPA are not expected to be any better than that reflected by the limits given, because no allowance is made in the above computations for the existence of any real differences (or biases) between EPA and the participating laboratories. In practice, an additional allowance, for some between-laboratory variability, based on previous survey results, is appropriately added (15).

It is helpful to present in graphical form the individual and combined effects of various sources of variability. The equation given previously for combining such effects assumes that the various effects are independent of each other--a valid assumption in many similar measurement systems. By making use of the Pythagorean Theorem, which relates the lengths of the sides of right triangles, the relative magnitudes of the standard deviations of the above expression (Equation 2) can be displayed graphically by the sides of a group of adjoining right triangles.

Thus, using the data from Table IV, the graphical representation for combining the within sample-day*, between day, and between-sample variations for each of the series is as shown in Figure 4.

Adding the between-day variabilities for series 1000 and for series 5000 gives the variabilities shown in Figure 4.

Thus, it can readily be seen that the between-day variability becomes an overwhelming portion of the combined variability for the higher concentrations. Further study should be initiated to identify the specific cause for the day-to-day variability.

It is of interest to compare the combined variability of the sampling and analytical variabilities with the variability experienced recently by participating laboratories analyzing these types of samples. Recent interlaboratory comparisons among laboratories result in computed coefficients of variation of 20% within the concentration ranges of this study.

The coefficient of variation of 20% includes the combined variation due to samples, days, and within sample-day, since each laboratory analyzed one sample (at each level) on some given day. Assuming that this combined variation for the participating laboratories is equal to that for the EPA laboratory, we must subtract this variability from the computed

---

*On Figure 4, the within sample-day variability, for the sake of space, is called "error", the usual statistical term for the inherent variability not identified with some specific factor.

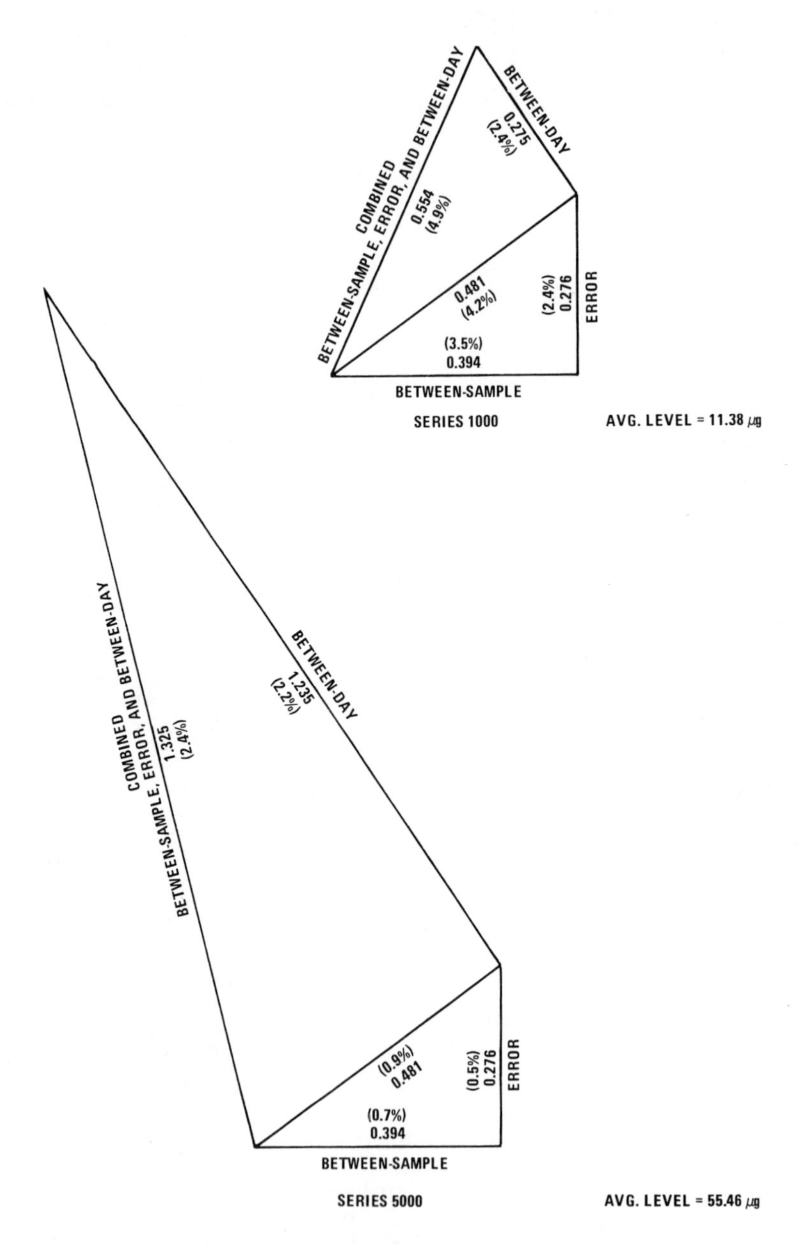

Figure 4. Combined between sample, within sample-day (error), and between day variabilities for series 1000 and 5000

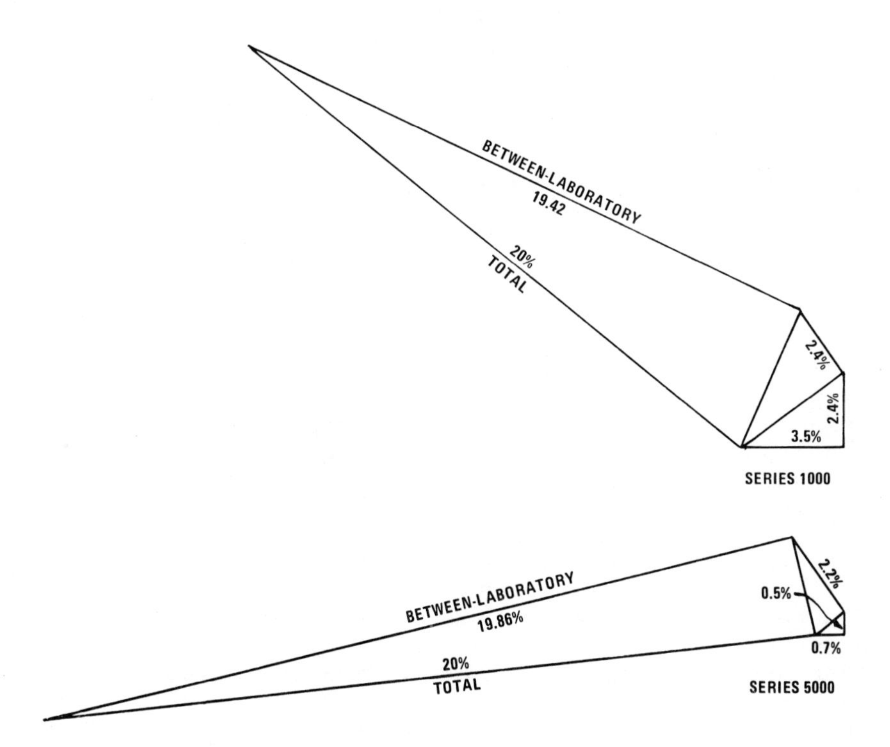

Figure 5. Combined within laboratory and between laboratory variability for series 1000 and 5000

20% value to obtain an estimate of the true variation between laboratories.

Addition of these between-laboratory components of variation to those of the previous figures provides the results shown in Figure 5.

Thus, if the participating laboratories are a representative sample of all the laboratories in the country, the graphical representation reflects the relative magnitudes of the sources of variation of the chemical analytical portion of SO_2 data being obtained throughout the country.

The picture clearly indicates that a fruitful area for investigation to improve the reproducibility of the reported data is the determination of the causes of the between-laboratory differences.

It should be noted that additional sources of variability, such as between analysts, could be considered in the planning and analysis for such studies, and that the graphical representation can also be extended to include these additional sources.

SUMMARY

Variability in the measurement process may be associated with instruments or apparatus, analysts, conditions, time, calibration sequence, calibration standards, and reagents and materials used in the entire measurement process. The entire laboratory measurement process has been viewed as a process requiring its own quality assurance system. Several pertinent points and means of control for achieving and maintaining good accuracy and precision of the product, DATA, of a laboratory measurement system have been presented. An example has been presented to show how designed studies of the various factors of the measurement process may be evaluated to determine the components of variance attributable to each of the factors. A graphical method of presenting the measures of individual and combined components of variability due to independent factors has been demonstrated. Much fruitful efforts in the study of laboratory measurement systems result from a close working relationship between chemists and statisticians.

LITERATURE CITED

1. Cameron, Joseph M., Journal of Quality Technology (1976) 8 (1) pp. 53-55, "Measurement Assurance".
2. Curley, James B., ASTM Standardization News, (1976) 4 (9) pp. 16-18, "Quality Assurance and Test Methology".
3. Wening, Robert J., ASTM Standardization News (1976) 4 (3) pp. 11-16, "Quality Assurance in the Laboratory".
4. Environmental Protection Agency, "Guidelines for Development of a Quality Assurance Program", EPA-R4-73-028 series and EPA-650/4-74005 series, Environmental Monitoring and Support Laboratory, (1973-1976).
5. Environmental Protection Agency, "Quality Assurance Handbook for Air Pollution Measurement Systems, Volume 1, Principles", EPA-600/9-76-005, Environmental Monitoring and Support Laboratory, Research Triangle Park, N.C., March 1976.
6. Von Lehmden, Darryl J., Raymond C. Rhodes and Seymour Hochheiser, "Applications of Quality Assurance in Major Air Pollution Monitoring Studies--CHAMP and RAMS," proceedings International Conference on Environmental Sensing and Assessment, Las Vegas, Nevada, September 14-19, 1975.
7. Cameron, Joseph M., Journal of Quality Technology (1975) 7 (4), pp. 153-195, "Traceability?".
8. American Society for Testing and Materials, "Manual for Conducting Interlaboratory Study of a Test Method", (STP 335), American Society for Testing and Materials, Philadelphia, PA, 19103.
9. Youden, William J. and Steiner, E. H., Association of Official Analytical Chemists (1975), "Statistical Manual of The Association of The Official Analytical Chemists, Statistical Techniques for Collaborative Tests, Planning and Analysis of Results of Collaborative Tests".
10. Lewis, Lynn L., ASTM Standardization News, (1976) 4 (9) pp. 19-23, "Interlaboratory Testing Programs for the Chemical Analysis of Metals".
11. ASTM E177, "Use of the Terms Precision and Accuracy as Applied to Measurement of a Property of a Material", American Society for Testing and Materials, Philadelphia, PA 19103.

12. ASTM E180-67, "Developing Precision Data on ASTM Methods for Analysis and Testing of Industrial Chemicals", American Society for Testing and Materials, Philadelphia, PA 19103.
13. Environmental Protection Agency, Federal Register (1971) 36 (84), pp. 8187-8190, "National Primary and Secondary Ambient Air Quality Standards".
14. Hicks, C. R., "Fundamental Concepts in the Design of Experiments", Holt, Rinehart, and Winston (1964).
15. Bromberg, S. M., Akland, G. G. and Bennett, B. I., "Survey of Laboratory Performance Analysis of Simulated Ambient Sulfur Dioxide Bubbler Samples", Environmental Protection Agency, Environmental Monitoring and Support Laboratory, Research Triangle Park, N.C. 1975.

INDEX

D

J

K

L

M

Q

R